高等院校土建学科双语教材（中英文对照）
◆ 土木工程专业 ◆
BASICS

砌 体 结 构
MASONRY CONSTRUCTION

[德] 尼尔斯·库默尔　编著

余流　译

中国建筑工业出版社

著作权合同登记图字：01-2007-3337号

图书在版编目（CIP）数据

砌体结构／（德）库默尔编著；余流译．—北京：中国建筑工业出版社，2011
高等院校土建学科双语教材（中英文对照）◆ 土木工程专业 ◆
ISBN 978-7-112-11600-3

Ⅰ．砌… Ⅱ．①库…②余… Ⅲ．砌块结构-高等学校-教材-汉、英 Ⅳ．TU36

中国版本图书馆CIP数据核字（2009）第210930号

Basics：Masonry Construction/Nils Kummer（Ed.）
Copyright © 2007 Birkhäuser Verlag AG（Verlag für Architektur），P. O. Box 133，4010 Basel，Switzerland
Chinese Translation Copyright © 2011 China Architecture & Building Press
All rights reserved.
本书经Birkhäuser Verlag AG出版社授权我社翻译出版

责任编辑：孙　炼
责任设计：郑秋菊
责任校对：赵　颖

高等院校土建学科双语教材（中英文对照）
◆ 土木工程专业 ◆

砌体结构

[德] 尼尔斯·库默尔　编著
余流　译

*

中国建筑工业出版社出版、发行（北京西郊百万庄）
各地新华书店、建筑书店经销
北京嘉泰利德公司制版
北京建筑工业印刷厂印刷

*

开本：880×1230毫米　1/32　印张：3⅞　字数：125千字
2011年5月第一版　2011年5月第一次印刷
定价：**16.00**元
ISBN 978-7-112-11600-3
（20280）

版权所有　翻印必究
如有印装质量问题，可寄本社退换
（邮政编码100037）

中文部分目录

\\ 前言　5

\\ 导言　71
　　\\ 砌体　71
　　\\ 砖　71
　　\\ 砌筑砂浆　71

\\ 施工工艺标准　72
　　\\ 尺寸和模数　73
　　\\ 计量尺寸及名称　76
　　\\ 砖块组砌方式　79
　　\\ 砌体砌筑　79
　　\\ 标准施工　83
　　\\ 灰缝构造　88
　　\\ 装修工艺标准　90
　　\\ 石材建筑　90
　　\\ 新方法　91

\\ 砌体结构　92
　　\\ 结构特性　92
　　\\ 外墙　93
　　\\ 内墙　105
　　\\ 凹槽和洞口　108

\\ 建筑材料　108
　　\\ 砌块类型　109
　　\\ 标准砌块　109
　　\\ 砌体砂浆类型　116

\\ 结论　118

\\ 附录　119
　　\\ 参照标准　119
　　\\ 参考文献　120
　　\\ 插图致谢　121

CONTENTS

\\Foreword _7

\\Introduction _8
    \\Masonry_8
    \\The brick_8
    \\Mortar for masonry_9

\\Rules of construction _11
    \\Dimensions and modules _11
    \\Unit dimensions and designations _15
    \\Brick courses _18
    \\Masonry bonds _18
    \\Regular constructions _22
    \\Joint configurations _29
    \\Finishing rules _30
    \\Building in stone _31
    \\New approaches _32

\\Masonry structures _35
    \\Structural behaviour _35
    \\External walls _36
    \\Internal walls _50
    \\Slots and gaps _54

\\Building materials _56
    \\Masonry unit types _56
    \\Standard masonry units _57
    \\Types of mortar for masonry _65

\\In conclusion _67

\\Appendix _68
    \\Standards _68
    \\Literature _69
    \\Picture credits _70

前　言

　　砖是一种主要的建筑材料。其形式简单，用途广泛；因地制宜，就地取材，来源丰富；砌块通过砂浆等材料粘结连接，节点形式灵活；同时砌筑的墙面表现丰富。但所有这些的前提是砌块材料必须满足一定的专业要求。

　　砌体作为建筑艺术的最普通的基本构成要素，路德维希·密斯·凡·德·罗（1886~1969年，20世纪最有影响力的艺术家之一，德绍包豪斯的最后一任校长）对其进行了高度赞美。然而砌体形式繁多，很难加以完整概括。从古代的歌剧院、巴比伦神庙到现代博物馆或者简单的居民住宅，都可看到它的身影。众所周知，没有砖和砂浆，对于建筑物来说是难以想像的。

　　砌体应用广泛，但要使其作用发挥得淋漓尽致，必须满足严格的专业技术标准。从砖到墙体、再到房间，最终到整个建筑的形成，整个过程并不容易。

　　本基础教材系列以适于实际应用的方式，将砌体专业有关知识向读者娓娓道来，培养他们的专业技能，内容深入浅出。本系列包括若干卷，每卷从一个基本点出发，然后加以深入探讨；引入一个主题，然后进行解释，给读者提供必要的实用专业知识，而不是对专业知识泛泛而谈。

　　本书向读者系统地介绍了砌体结构。砖和砂浆作为基本的组成要素，常用来砌筑墙体。砌体体系整体构成和材料性能对于理解砌体墙体特性至关重要。本书从砖到墙体，对砌块相互作用、砌体砌筑形式、带缝隙、凹槽和透视投影效果的砌体墙体美学效果均进行了绘声绘色的系统阐述，使读者能够理解、掌握砌体的精髓，并将这种理解直接应用于设计和工程实践。

丛书主编：贝尔特·比勒费尔德

FOREWORD

"The brick is another master-teacher. How profound that little format is, handy, how useful for every purpose. What logic its structure shows in bond. How lively is that play of joints. And what richness even the simplest area of wall possesses. But what discipline this material demands."

What Ludwig Mies van der Rohe (1886–1969), one of the most influential artists of the 20th century and the last director of the Dessau Bauhaus, is enthusiastically celebrating here is nothing other than one of the lowest common denominators and at the same time essential basic elements of any architecture: masonry. It appears in so many different forms that it is scarcely possible to provide a complete survey. Whether you look at ancient amphitheatres, Babylonian temples, modern museums or simple houses: without bare brick combined with simple mortar, architecture as we know it today can scarcely be explained.

But the well-nigh infinite creative variety afforded by masonry conceals strict rules that have to be obeyed if the desired overall impression is to be guaranteed. The pathway from the brick to the wall, to the room and finally to the whole building is neither short nor simple.

The "Basics" series of books aims to present information didactically and in a form appropriate to practice. It will introduce students to the various specialist fields of training in architecture. Content is developed stage by stage, using readily understandable introductions and explanations. The essential points of departure are built up systematically and explored further in the individual volumes. The concept is not to provide a comprehensive collection of expert knowledge, but to introduce the subject, explain it, and provide the necessary expertise for skilled implementation.

The present volume aims to introduce students systematically to the subject of masonry. Bricks and mortar, the elemental basic components, are used to devise rules for building a wall. The emphasis is on the overall systems and material-dependent properties that are essential for understanding a "wall". The interplay of bricks, the forms of masonry bonds, and the aesthetic of masonry walls with apertures, projections and recesses are explained soundly and methodically – from brick to wall – so that students can understand the essence of masonry and apply their insights directly to their designs and projects.

Bert Bielefeld
Editor

INTRODUCTION

Masonry buildings cannot be reduced to any particular tradition, fashion or style: timeless in their flexibility, fundamental to both classical and avant-garde architecture, open to stylistic trends throughout the ages, capable of being both ordinary and experimental. The façades of contemporary high-rise buildings and modern glass structures may manage without classical masonry as a basic architectural principle, but it is difficult to find buildings without a masonry wall somewhere inside, thus reconfirming the existential character of masonry.

The book will present the "Basics" of masonry, together with the demands it makes. First of all we need to look at its basic components, bricks and mortar. The rules for fitting these elements together to make a wall form the theoretical and creative basis in the second chapter for understanding the wall constructions explained in the third. Then, moving from plain brick construction to the completed wall, we arrive in the fourth chapter at the question of which building materials are suitable for the types of work described earlier, thus ending up with the brick again.

MASONRY

Masonry is not a building material like wood or steel, but a combination of two individual materials, bricks and mortar, handled according to the rules of a craft. It is frequently classified as a composite material, and thus more like reinforced concrete than concrete, for example, as the quality of the end product depends on the quality of both the materials and the execution.

Masonry is used primarily for constructing walls, as a loadbearing or partitioning shear wall, as facing to protect or clad, or as infilling between columns and beams. Bricks are also found in vaults or coping, and also as a floor covering.

There are different kinds of bricks and mortar for all these functions and structures. It is therefore important to know the most important properties of the two materials as well as about construction, so that the ideal combination can be achieved.

THE BRICK

Asked about bricks, most people would probably sketch a uniform shape and size: the standard brick. Developed over millennia, bricks lend

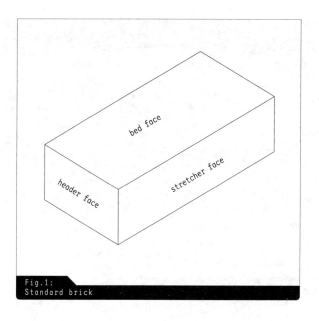

Fig.1:
Standard brick

their character to most masonry façades and are firmly linked with our idea of masonry. But there are a whole variety of different shapes and sizes: flat Roman bricks, large manufactured blocks or octagonal moulded bricks. And the brick forms the basis for the whole set of craft rules of erecting a masonry building, even today. These rules govern the size of rooms and buildings, apertures and built-in features, and they structure façades.

MORTAR FOR MASONRY

The second component of masonry is mortar. It enables the bricks to cover a full area, balances tolerances and ensures that the bricks will hold together strongly, and its finish and colouring influence the look of exposed masonry. It is applied both horizontally between the individual layers of bricks (course joint) and vertically between the individual bricks (perpend). Even though modern manufacturing methods are shrinking the layers of mortar for reasons of cost and structural engineering technology, the combination of brick and mortar is crucial when planning construction. The cohesion of mortar and brick, and thus also the choice of individual components, are important in terms of loadbearing capacity, so that even modern building methods without mortar follow rules based on traditional building.

RULES OF CONSTRUCTION

As masonry is a craft, there are certain rules for achieving a high-quality finish. The most important aims are:

_ Optimizing the loadbearing and resistance properties of the construction
_ Minimizing loss of material
_ Speeding up the building process
_ Executing a design that does justice to material and use

These rules form a theoretical basis for the wall structures given in the third chapter. They show the principles and methods for creating masonry from its components, the preferred dimensions, and how to form connections and apertures correctly for the material. The individual wall will be considered first.

DIMENSIONS AND MODULES

One of the architect's main tasks when planning and constructing a building is to coordinate and combine the various structural and craft services. Shell construction (walls, columns, floors etc.) and finishing (windows, doors, wall and floor coverings etc.) have to be matched in order to build efficiently. The actual building process, as well as planning and finishing, are simplified by repeating elements and sizes. However, fixing grid dimensions is problematical for masonry, as it is impossible to work with the dimensions of the bricks alone, since we must allow for the mortar joints between the bricks as well. Here, a simple device is used to decide when the joint must be added to a wall length or not: the distinction between specified and nominal dimensions.

<small>Specified dimension and nominal dimension</small>

The specified dimension is the basic theoretical measurement, the grid or the module multiplied to put together the whole masonry construction system. The nominal dimension, however, is the dimension that is actually executed and entered on the construction drawing. This distinction can be used to systematize construction types with joints, and particularly masonry constructions.

Although the nominal and specified dimensions are identical when building without joints, they are treated as follows in construction for building types with joints:

The specified dimension consists of the nominal dimension executed and the corresponding joint, i.e.:

brick dimension + joint

Conversely, the nominal dimension is the brick dimension alone, without joint.

If you now imagine a masonry wall with window apertures and transverse walls, you will very quickly realize that there are different dimensions for the wall width, aperture and projections because of the mortar joints.

External dimension

The external dimension is the wall thickness. One joint must always be subtracted from the specified dimension as there is always one joint missing, regardless of the number of bricks.

External dimension (E) = specified dimension – joint

Aperture dimension

The interior dimension of an aperture always contains an additional joint.

Aperture dimension (A) = specified dimension + joint

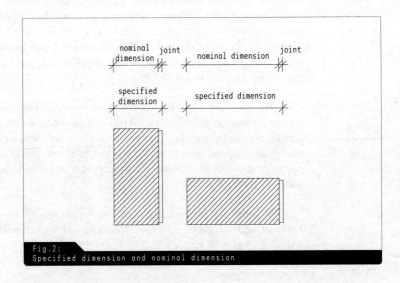

Fig.2:
Specified dimension and nominal dimension

Projection dimension

The projection dimension measures the piece of wall between opening and wall or wall projections. Here, the missing joint in the external dimension and the additional joint in the aperture dimensions balance each other out.

Projection dimension (A) = specified dimension

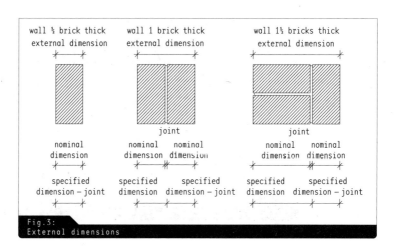

Fig. 3:
External dimensions

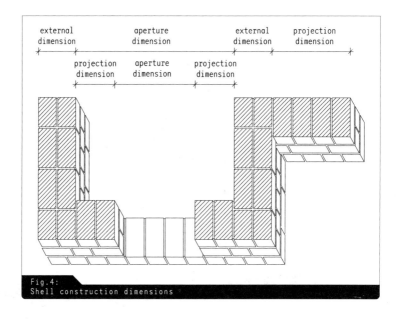

Fig. 4:
Shell construction dimensions

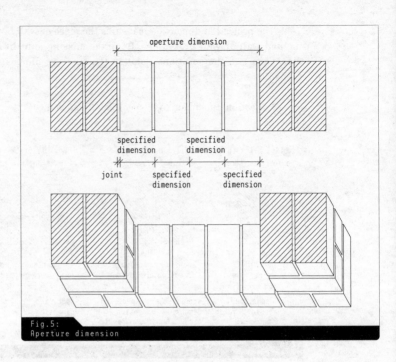

Fig.5:
Aperture dimension

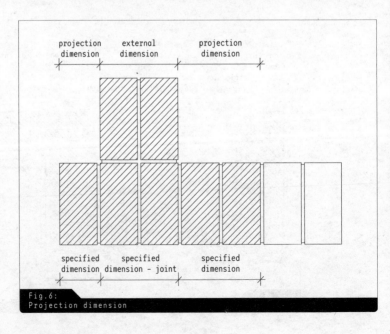

Fig.6:
Projection dimension

UNIT DIMENSIONS AND DESIGNATIONS

These hitherto theoretical definitions have left open the question of actual dimensions, which are independent of the brick and joint sizes chosen. These sizes can vary, and have led to different standards in different countries, according to local traditions.

In Germany, masonry is based almost exclusively on the octametric system, which uses an eighth of a metre = 12.5 cm as the specified dimension. The standard brick, so-called "normal format", measures 24 × 11 × 7.1 cm (nominal dimensions). When the joint sizes for of 1 cm for the vertical head joints and 1.23 cm for the horizontal course joints are added, this gives specified dimensions of 25 × 12.5 × 8.33 cm, multiples of which produce a metre.

Joint sizes can also vary, without changing the system. New manufacturing technology and the need to meet the greater-than-ever demands on masonry for heat and sound insulation, and in terms of loadbearing capacity, mean that masonry technique is no longer based on the centimetre joint. Modern manufactured blocks are finished to such low tolerances that joints need be only a few millimetres thick.

However, to maintain the usual specified dimensions, the unit dimensions have been adapted to ensure that the overall dimensions still fit in with the system:

For example:

Traditional:	German normal format brick	24 cm + 1 cm joint = 25 cm
Modern technique:	Manufactured block	24.7 cm + 3 mm joint = 25 cm

\\Tip:
In Germany, these dimensions are fixed by the DIN 4172 standard dimension in the building industry, which has prescribed a basic module of 25 cm for shell construction since the post-Second World War rebuilding, basing itself on traditional formats. The later DIN 18000 modular standard for building, which promised to be simpler to use with its decimetric basic module of M – 10 cm, has not caught on in Germany.

\\Hint:
Different countries have other standard bricks, based on national traditions or different units (e.g. inches), e.g. 21.5 × 10.25 × 6.5 cm in England, 19 × 9 × 6.5 cm in Belgium, and 8 × 4 × 2.25 inches (20.3 × 10.2 × 5.7 cm) in the USA.

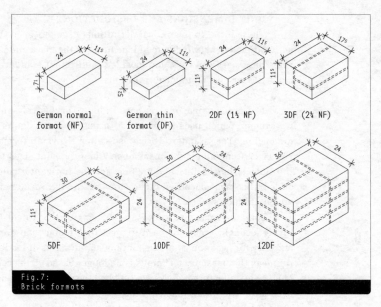

Fig.7:
Brick formats

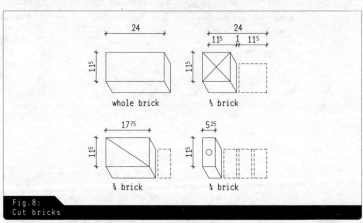

Fig.8:
Cut bricks

\\Hint:
As the same number of thin-format units can be combined in different ways, different formats produce the same designation, e.g. 8DF = 24 × 24 × 23.8 cm and 8DF = 24 × 49 × 11.3 cm.

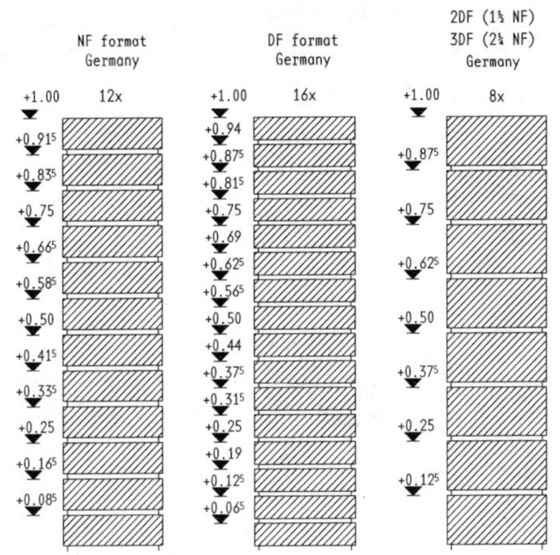

Fig. 9: Height comparison

Small formats are also distinguished:

L × W × H = 24 × 11.5 × 7.1 cm – normal format (NF)
24 × 11.5 × 5.2 cm – thin format (DF)

Larger bricks are made up of several thin formats as a basic module with the corresponding joints, and are thus defined as 5DF, for example.

Examples for dimensions in the octametric system:

Specified dimensions: 12.5 cm; 25 cm; 37.5 cm; 50 cm ... 100 cm etc.
Nominal dimensions: 11.5 cm; 24 cm; 36.5 cm; 49 cm ... 99 cm etc.
External dimensions: 11.5 cm; 24 cm; 36.5 cm etc.
Aperture dimensions: 51 cm; 1.01 m; 1.26 m etc.
Projection dimensions: 12.5 cm; 25 cm; 1.00 m etc.

When bricks are cut, always remember to subtract a joint:

¾ brick = specified dimension/4 × 3 – joint = 6.25 cm × 3 – 1 cm
= 17.75 cm.

Cut units are specially designated in the top view on laying drawings: the ¾ unit (17.75 cm) by a diagonal, the ½ unit (11.5 cm) by a cross, and the ¼ unit (5.25 cm) by a point or a circle.

The octametric numerical values are used for height as well. To achieve the specified dimension height (25 cm, 50 cm, 1 m etc.), the horizontal mortar joints serve as a height levelling course, and thus measure between 1.05 and 1.22 cm.

BRICK COURSES

The individual rows in a masonry structure are called courses. A distinction is made according to the run of the bricks:

Stretcher course:	bricks are laid parallel with the axis of the wall
Header course:	bricks are laid transversely to the wall axis
Brick-on-edge course:	bricks are laid transversely and standing edgewise on their long sides
Soldier course:	bricks stand edgewise on their narrow sides as an upright header course

While the stretcher and header courses are combined with each other in different ways as bonds, the edge and soldier courses with their larger head joints offer greater bond strength between the bricks and better pressure dispersal, as they do not break as easily as a horizontal brick. They are therefore used for lintels, seatings and cornices.

MASONRY BONDS

To produce high-quality masonry with a good loadbearing capacity from bricks and mortar there are certain craft rules that must be followed when laying bricks – the bond rules. These rules distinguish between four so-called school bonds – those most commonly taught – according to the sequence in which the brick courses are laid on one another, and the way they are offset from each other.

Some of the bond rules are general, and give us the first two school bonds.

Rules:
_ All the courses must be laid horizontally.
_ The brick height should not be greater than the brick width.

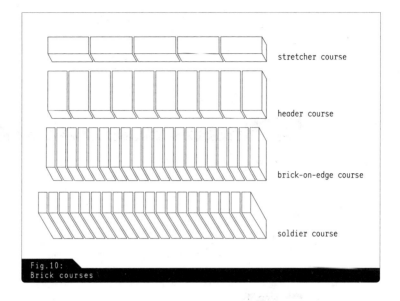

Fig.10:
Brick courses

_ Only bricks of the same height should be used in a single course (only at wall ends can there be exceptions in every second course).
_ The largest possible number of whole bricks should be used.
_ The offset between the courses is at least ¼ brick length for all perpends.

The brick offset is crucial for the wall's loadbearing capacity. The greater the offset, i.e. the more shallow the racking back of the bricks, the greater the resistance to longitudinal cracks.

\\Hint:
The designation "brick length" also relates to the corresponding standard brick in terms of brick offsetting. But the joint must always be taken into account. Thus, with a specified dimension of 25 cm: ¼ brick length = specified dimension/4 - joint = 25 cm/4 - 1 cm = 5.25 cm. The same applies to the brick width or the wall thickness: a wall two bricks thick = 2 × 25 cm - 1 cm = 49 cm (external dimension).

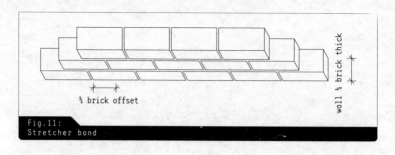

Fig.11:
Stretcher bond

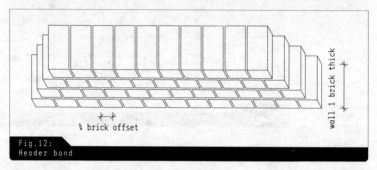

Fig.12:
Header bond

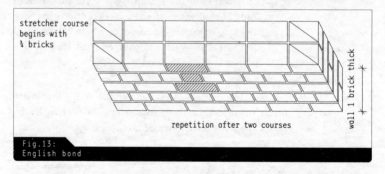

Fig.13:
English bond

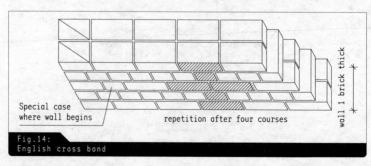

Fig.14:
English cross bond

Stretcher bond   In stretcher bond, all courses in the masonry are made up of stretcher courses offset by the length of 1/2 brick. As this bond does not permit an offset running transversely to the wall axis, it can only be used for wall ½ brick thick, e.g. for internal walls, facer skins and chimneys. A wider wall can be built only with larger bricks. Stretcher bond offers good compressive and tensile strength because of the large brick offset. It is also possible to use an offset of $1/3$ or ¼ of a brick length, but this entails some loss of loadbearing capacity.

Header bond   In header bond, all the courses consist of header courses offset by the length of ¼ brick. This bond can only be used for one-brick walls. Because of the low overlap the bond has less loadbearing capacity and inclines to diagonal cracks because of the steep racking. It is however particularly suitable for narrow masonry radii.

Combining these bonds and following two more rules gives the last two commonly taught bonds.

Rules:
_ Stretcher and header courses alternate.
_ One stretcher course begins with a ¾ brick (for thicker walls, with correspondingly more ¾ bricks).

English bond   English bond consist of alternate courses of stretchers and headers. The offset is ¼ brick. This produces usefully shallow racking by ¼ and ¾ brick lengths in each case.

English cross bond   Like English bond, English cross bond begins with alternate stretcher and header courses. But the perpends of the stretcher courses are offset against each other by ½ brick, so the joint pattern repeats only every four courses. This bond has a more varied joint pattern, but it is also more steeply racked and therefore more prone to diagonal cracks.

There are also some decorative bonds, but these are only of historical or regional significance. Examples are double Flemish bond, Yorkshire bond and Flemish bond.

It is also possible to achieve a lateral offset within the wall, and so to construct walls with a thickness greater than one brick, by alternating stretcher and header courses.

Additional rules apply here:
_ Only headers should be used if possible for wide walls.

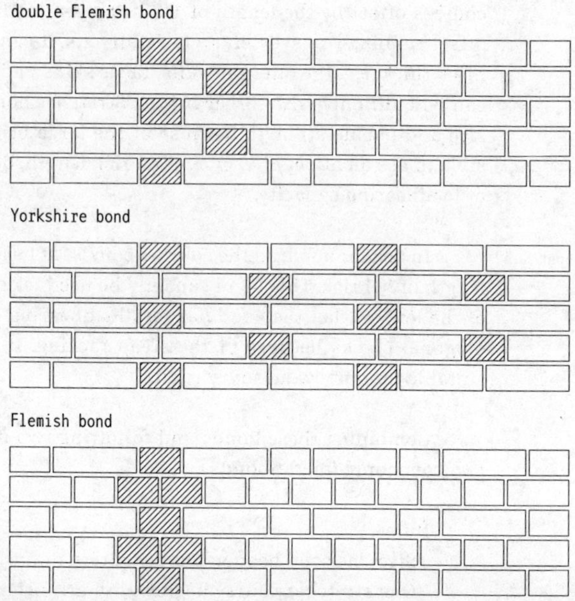

double Flemish bond

Yorkshire bond

Flemish bond

Fig.15:
Historical bonds

_ Perpends should run through the total thickness of the wall if possible.
_ The offset should also be at least ¼ brick length as the courses rise for the intermediate joints as well (perpends inside the wall).
_ The offset must be maintained longitudinally and horizontally.

REGULAR CONSTRUCTIONS

Corners in walls

For corners, niches, projections and columns, there are special points of detail covered by the bond rules.

Rules:
_ The stretcher courses run through at corners, junctions and joints; the header courses abut.
_ Parallel walls should have the same sequence of courses.
_ Only one perpend in each course should start from an inside corner.

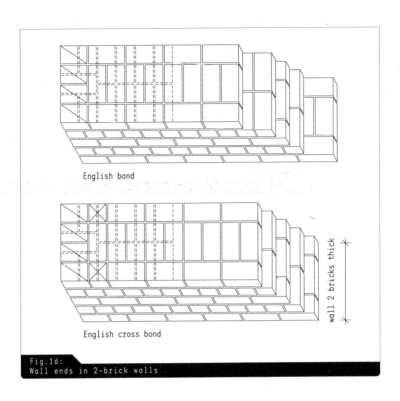

Fig.16:
Wall ends in 2-brick walls

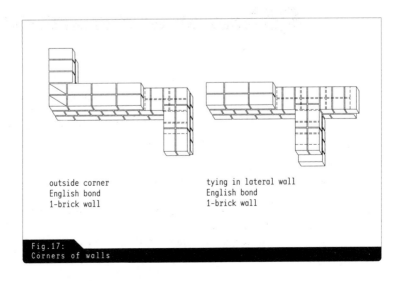

Fig.17:
Corners of walls

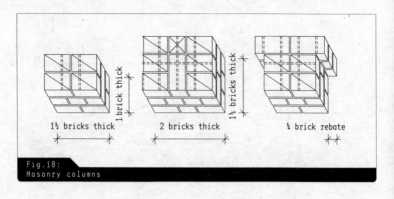

Fig.18:
Masonry columns

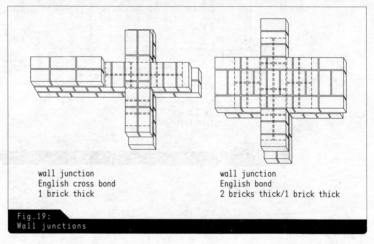

wall junction
English cross bond
1 brick thick

wall junction
English bond
2 bricks thick/1 brick thick

Fig.19:
Wall junctions

_ Windows and door strips should be constructed like wall ends with projections – for the headers by displacing one brick in the direction of the projection, for the stretchers by advancing the stretchers.

Masonry columns

Two points should be noted when constructing masonry columns:

_ Square columns have the same bond in every course, turned through 90° each time.
_ Rectangular columns start with ¾ bricks on the narrow sides, like wall ends. The gap is filled with whole or half bricks.

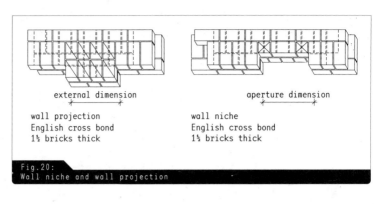

wall projection
English cross bond
1½ bricks thick

wall niche
English cross bond
1½ bricks thick

Fig.20:
Wall niche and wall projection

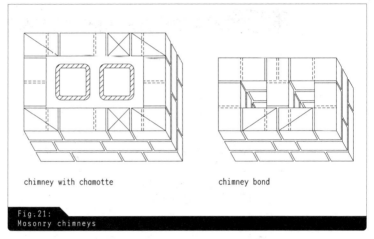

chimney with chamotte

chimney bond

Fig.21:
Masonry chimneys

\\Hint:
Because of new brick formats and techniques for constructing loadbearing walls, which are generally built using "random masonry bond" (not following the rules of bonding, but keeping to the standard minimum dimensions for the offset), these school bonds are generally used only for exposed masonry (see chapter Masonry structures, External walls).

\\Hint:
Chimneys are almost always built of special-purpose bricks today, so the exposed masonry structure shown here merely illustrates the possibilities and rules of bonded masonry.

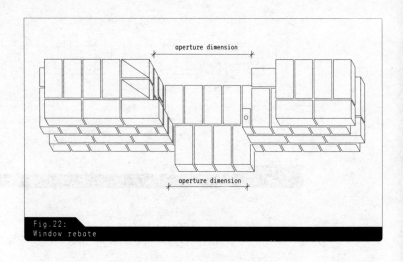

Fig.22:
Window rebate

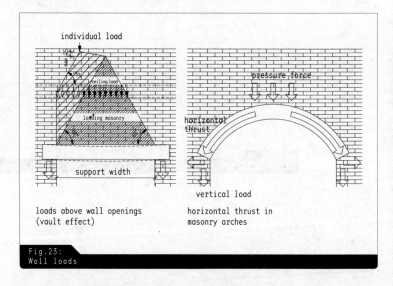

Fig.23:
Wall loads

Masonry apertures

Apertures for windows, doors or passageways in the wall are subject to craft rules and traditions, as well as the wall itself.

Side rebates for windows and doors can be constructed according to bonding rules; this simplifies installation and improves the fittings' resistance to rain and wind.

> ◐

Masonry arches

The top of the door or window can also be built according to the rules. As masonry cannot absorb bending forces, apertures cannot be topped with bonded masonry without "support", so beams, formerly made of wood or stone, and now of concrete, can be placed over the aperture. The beams dissipate the imposed load from the masonry above into the side walls through the structural conditions in terms of bending, restricting the possible size of the aperture according to the material used for the beam.

Another aperture suitable for masonry is an arch with masonry above it, which transforms all the imposed loads into pressure forces and transfers them to their points of support. The difficulty of this construction lies in the horizontal thrust that the loaded arch exerts on the masonry. This thrust, which increases in shallower arches, must be absorbed either by the wall or by additional piers.

Round arches are semicircles of masonry that transfer the imposed load into support points, which are usually horizontal. The radius of the arch is thus half the width of the aperture and lies at its midpoint. To achieve this radius the joints between the bricks should be wedge-shaped. A thickness of at least 5 mm may be reached on the inside of the arch (intrados) and a maximum of 20 cm at the other extremity (extrados). This means that when dealing with larger radii and aperture widths, several rows of bricks must be placed on top of each other. Wedge-shaped bricks can also be used for tighter radii.

If the radius is increased to the full width of the aperture and circles are drawn around the two support points, a pointed arch is produced. Both types of arch should consist of an uneven number of bricks, so that a keystone, which starts the load distribution, can be placed at the apex of the arch, rather than a perpend. The keystone should end in a bed joint of the masonry, so that the filler courses above the apex of the arch do not

◐
\\ Hint:
Because of the so-called "vault effect" of the masonry, which transfers the loads around the aperture, only the self-weight of the masonry above the apertures affects the beam, relating to a triangular load take-up area. In addition there are single loads, provided that they are not more than 25 cm above the tip of the take-up areas, and ceiling loads, if they are within the take-up area (see Fig. 23).

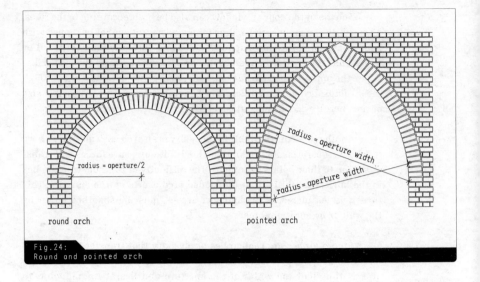

Fig. 24:
Round and pointed arch

become too large. For window rebates, arches can be built in two rows of bricks, displaced vertically.

If the surrounding loadbearing structure is able to absorb greater horizontal forces, a shallower arch structure may be chosen. For a segmental arch a circular sector with a greater radius is built; here the rise of the arch (the difference in height between the lowest and the highest point of the inside of the arch) must not be greater than 1/12 of the aperture width. The support points are tilted to point towards the centre of the arch.

If the aperture is built over almost horizontally as a result of the sideways tilt of the bricks, the term "straight arch" is used. Here the rise is reduced to a maximum of 1/50 of the aperture width.

The aperture width is strictly limited for both these construction methods. The following formula can be used as a rule of thumb:

_ 1.2 m for segmental arch with bricks 24 cm high
_ 0.8 m for straight arch with bricks 25 cm high

Masonry arches are very elaborate structures, commonly associated with churches and prestigious buildings, and are now only rarely built. Today the arches can be manufactured with steel reinforcements and built in as finished parts.

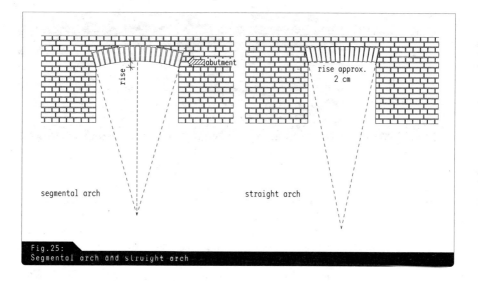

Fig.25:
Segmental arch and straight arch

JOINT CONFIGURATIONS

In addition to masonry bonds, the configuration of the mortar joints can make a considerable contribution to the appearance of the masonry. The colour or depth of the joints can emphasize them or make them inconspicuous for design purposes.

Executing joints correctly also makes the structure more resistant, and helps it to last longer. There are two kinds of joint:

Flush pointing

For trowel-finished joints the mortar pushed out at the sides when a brick is put in place is struck off and smoothed down a little later with a piece of wood or a hose. The advantage of this method lies in the good seal it creates for the joint and the need to apply the mortar to the whole surface, which improves the loadbearing capacity of the masonry.

\\ Example:
Frank Lloyd Wright emphasized the horizontal orientation of his Robie House in Chicago by recessing the bed joints and using flush perpends.

Fig.26: Joints

Subsequent pointing

However, if the uniformity of the joints is important in terms of colour and design, it can be advantageous to point subsequently. Here, the fresh mortar is scratched out with a wooden lath to a depth of about 20 mm and the opening cleaned; if absorbent bricks, which draw the water out of the mortar, are being used, the opening must be moistened before being closed again with the pointing mortar. Attention must be paid to high-quality finish because of the two kinds of mortar; this will guarantee the loadbearing capacity and density of the construction.

FINISHING RULES

Masonry must be bonded, and also built horizontal, true and plumb. The first course is crucial, as it compensates for uneven terrain. The rows of bricks should then be laid from the corners. This can be done by hand up to a weight of 25 kg per brick, above which auxiliary equipment is needed. The mortar must be applied to the full area of the bed joints; for small bricks with a trowel, for larger sizes with a mortar template, which keeps the height of the joints consistent over the full length of the wall. The perpends must also be closed to ensure that the masonry is rain- and wind-proof, either by covering the full surface with mortar or by flushing mortar pockets in the middle of the brick. To save time and expense, masonry is sometimes built without using mortar for the perpends. But here it is essential to meet all the demands of weather protection (by using a layer of rendering or cladding) and sound insulation (good sound-reducing bricks). Bricks using a tongue and groove system are preferred for this.

If highly absorbent bricks are being used, care should be taken before laying them to dampen the wall, as the bricks will draw too much water out of the drying mortar. The bricks will then also absorb fewer salts from the mortar that would later be visible on the surface of the brick as

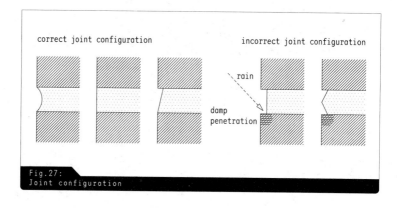

Fig. 27:
Joint configuration

"efflorescence". At the same time, a completely soaked brick will prevent proper binding with the mortar. Bricks and masonry should therefore be protected against rain, as well as against unduly strong sunlight. In frost, bricks can only be laid if precautions are taken, as mortar hardens more slowly with falling temperatures, and stops hardening altogether at −10 °C. Building materials should be covered as soon as the temperature falls to 5 °C, and at temperatures under 0 °C bricks and mixing water should be warmed. Frozen materials should not be used, and parts of the wall that have already been damaged must be removed.

BUILDING IN STONE

Natural stone is the Ur-form of masonry. From the simple, mortarless piling up of unworked stones in various sizes (drystone walling) to stones of equal sizes laid according to bonding rules (ashlar masonry), there are various special types of natural stone masonry. However, natural stones are now used less for actual masonry than as curtain façade material for

\\ Example:
Arno Lederer chose this colour design option for his office building in Stuttgart. He used a black brick and pointed the perpends in black as well, but the bed joints are white. This gives the façade an unmistakable appearance (see Fig. 26, second picture from the left).

> **\\Hint:**
> Thermal bridges are weak points that cause heat loss from a building. They can be determined, geometrically if the areas absorbing heat are smaller than those giving it out (e.g. at the corners of buildings); by the material, if different materials are used; or structurally by heat-conducting fastenings and penetrations.

walls, and therefore they will not be dealt with more fully here. Exceptions are primarily found in monument protection and landscape architecture.

NEW APPROACHES

In addition to the traditional building method prescribed by the bonding rules, new approaches have developed, based on new manufacturing methods and building materials, intended above all to make masonry construction cheaper and less time-consuming.

Moulded brick masonry

For moulded brick masonry, the dimension tolerance of the bricks has been minimized so that the joint height can be reduced to 1–3 mm (thin bed). The mortar is applied with a roller, or the bricks are dipped in the water. As the joint proportion is minimized and homogeneous masonry produced, material and time are saved, and favourable statical values achieved, > see chapter Masonry structures, Structural behaviour and there are fewer thermal bridges.

As the brick rows can only accommodate low tolerances, the first layer should be laid with great care. Small offset blocks can be used for this purpose. They are available in different heights and with good insulation properties.

Dry masonry

Dry masonry uses no mortar at all. For reasons of loadbearing capacity, however, such wall constructions are restricted to low storey and building heights. The ceiling loads on the walls must be even, so that this pressure can compensate for the lack of adhesion from the mortar.

Masonry kits

To save the time needed for cutting large stones to size, masonry kits offer the possibility of assembling whole sections of walls in the right dimensions in the factory, and delivering them to the site as individual parts with a laying plan. This method is a reasonably priced alternative, particularly if there are many diagonals (gable walls) or apertures.

Prefabrication construction method

This method takes prefabrication a little further at the factory stage: manufacturers deliver whole storey-height walls, including apertures, to the site. The bricks have to be reinforced to stabilize the structure and erection requires a crane or mobile crane. The expense is set off against the consistent quality of the factory work (although the erector is of course responsible for the wall connection points).

MASONRY STRUCTURES

The structures listed below refer to the wall in its built state. The construction rules explained above apply in principle to all masonry structures, and deal merely with assembling bricks and mortar. There are various ways of finishing a construction, combinations with other building materials and dependencies on other parts of a building. These relate to the location where the building is to be used and the role of the wall structures.

Masonry walls can be loaded vertically from ceilings and other parts of the building, by self-weight and also by horizontal forces such as wind, soil pressure and impact forces, or cantilever loads from projecting or suspended elements.

For these reasons the walls must be connected non-positively with the adjacent building parts, i.e. the loads must be transferred via other loadbearing sections or directly into the foundations. The wall is stabilized by tie walls that prevent buckling, and by even vertical loading. When dimensioning these walls, there are more requirements relating to the building science of fire protection to meet. Walls supporting nothing more than their own weight from one floor and forces occurring horizontally to the wall level can also be built as non-loadbearing.

STRUCTURAL BEHAVIOUR

The loadbearing capacity of masonry is determined by the bonding of brick and mortar. The adhesion or friction between brick and mortar affect how horizontal forces are absorbed and provide vertical load distribution over the full area; the joint compensates for brick tolerances. Its ability to absorb compressive forces is far greater than its acceptance of tensile or tensile bending forces, i.e. a precisely bedded brick can transfer loads as compressive force but would break without this bedding surface. The brick and its mortar joint are pressed together from above in vertical loading. The brick conducts the compressive forces, and the mortar, which is more easily deformed, tries to sag. These different behaviours produce stress at the point of contact between bricks and mortar, and then to compressive stress in the mortar and tensile stress in the brick. At the same time, this lateral tensile stress in the brick reduces its compressive strength. If the load becomes too great, vertical cracks will appear in the brick and the mortar will collapse. Uneven application of mortar increases tension peaks and the danger of collapse. Greater join thicknesses and the use of

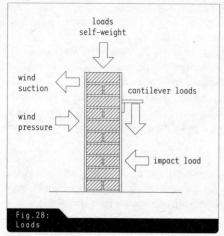

Fig.28: Loads

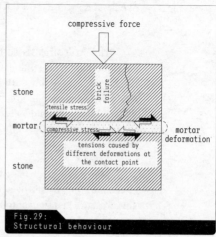

Fig.29: Structural behaviour

lightweight mortar are also hazardous because of their greater deformability. Heavy bricks with a high specific density transfer forces well.

Aerated bricks and cavities weaken the cross-section and thus loadbearing capacity. Adhesion between bricks and mortar also allows force to be absorbed horizontally.

The brick's compressive strength is another crucial factor. Brick and mortar must also be matched to each other to avoid the joint collapsing. Compressive strength classes are given as characteristic values for both bricks and mortar.

EXTERNAL WALLS

External masonry walls are loadbearing walls, except for infill within other loadbearing systems (frame construction, construction slabs

> \\Hint:
> Specific density is the ratio of mass to volume. As it is increased by absorbing water, this value is usually given for dry bricks, the dry density, in kg/m³.

etc.), or free-standing walls. They also separate the inside of the building from the outside, and so must give protection against cold, rain, snow and sound from the outside. At the same time, design questions play a part in decisions about whether the masonry should be visible from the outside or not.

Single-leaf masonry

External walls with just one wall built in bond are called single-leaf masonry. This structure, simple to erect in terms of craftsmanship, has to perform all the functions of an outside wall.

Single-leaf exposed masonry

Single-leaf exposed masonry, a wall structure that is visible from both sides or at least from the outside, displays a disparity between thermal insulation and weather protection. In order to meet today's thermal insulation criteria, aerated bricks providing offering good insulation must be used. As still air has a very low capacity for specific thermal heat conductivity and very low density, bricks with a high proportion of air in the form of pores or cavities, and thus a low specific density, provide good thermal insulation, but at the same time scarcely any protection against weather.

Their pores quickly become permeated with moisture, they are not frost-resistant and thus not suitable for unprotected use. Conversely, weather-resistant bricks with a high specific density offer little resistance to heat penetration and would require uneconomic wall thicknesses. This structure can therefore no longer be used in this way.

Facing masonry

For facing masonry, on the other hand, a wall several units thick between two different kinds of masonry is used inside the bond, so that the bricks showing on the outside offer good protection against weather and frost, and the inner series takes over the thermal insulation. Here the whole cross-section including the facing can be added to the load dispersal;

\\Hint:
The specific thermal conductivity (λ) indicates how much heat a structural element will transfer under fixed conditions. The smaller the value, the better the thermal insulation.

\\Hint:
The thermal transfer resistance (R) indicates a structural element's insulation capacity, according to its thickness. It is calculated from the ratio of course thickness to specific heat conductivity. The transitions at the extremities of the element are also calculated, and the individual values added for multi-course elements.

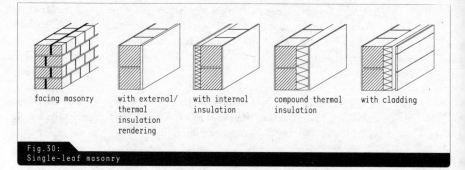

Fig.30:
Single-leaf masonry

the brick with the lowest compressive strength provides the basis for calculations. A joint between the two series of stones, offset course-wise, 2 cm thick and closed with seal mortar, offers protection for the inner set of units. This is an elaborate structure, and the units in it must be well matched to each other, in order to avoid different settling rates and deformations. Very precise planning is also needed, because unit formats often differ. This structure is recommended only for visual or formal reasons, or if a special brick is to be used or there is a request to manage without expansion joints in the exposed masonry. > see chapter External walls

Because of all these interdependent features, additional measures have to be taken with single-leaf masonry to protect it from the weather.

Single-leaf masonry with external rendering

Thus, for example, external rendering can be applied; this improves thermal insulation as thermal insulation rendering. The visual effect of bonded masonry is lost in single-leaf masonry with external rendering, but large-format units can be used, built in random bond with a thin mortar bed. They have better insulation properties, and are economical to use. As the whole cross-section of the wall contributes to the thermal insulation, weaknesses must be avoided to prevent thermal bridges. Special constructions are needed, especially for lintels and ceiling supports.

Ceiling supports
>

Ceilings must be connected with the enclosing walls by their supports on the wall via adhesions and friction. As a rule a bearing edge of 10–12 cm is needed.

>

As reinforced concrete has a lower heat transfer resistance than masonry, full support for the ceiling reduces thermal insulation. This produces colder ceiling and wall areas, and moisture from warm interior air may condense on their inside faces.

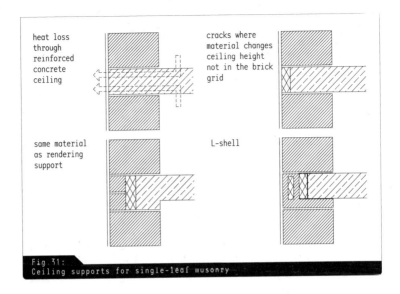

Fig. 31:
Ceiling supports for single-leaf masonry

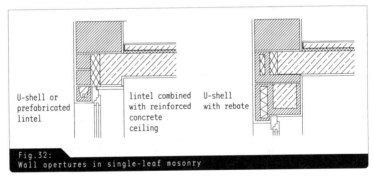

Fig. 32:
Wall apertures in single-leaf masonry

\\Hint:
If there is not enough area to support the ceiling, steel tie bars must be fixed into the masonry. As this exposes the masonry to horizontal tensile forces, the wall areas must receive a corresponding imposed load to counter the tensile forces. This means that ties cannot be fixed in parapet areas. Gable walls can also be attached to the roof structure with tie bars.

\\Example:
Condensate: Warm air can hold more water vapour than cold air. If warm air meets cold air, mist or water vapour is formed. If warm air meets a cold object, excess water is released, forming a condensate. For heated air in a room, if an outside wall is poorly insulated, or even not insulated at all, water will be deposited on the cold internal side or in the cooled structural element. This then leads to structural damage from frost or mould.

For this reason, additional insulation must be provided at the outer edge of the ceiling. It should be noted here that single-leaf rendered walls may be subject to cracks on the outside because of different expansion and deformation at the point where the different materials meet, so that rain may penetrate the building. Fabric can be applied to bridge the point of transition and secure the rendering, but the use of L-shells is also recommended. These are made of the same material as the wall, and some already have insulation strips. They avoid the change of material while acting as a formwork element for the reinforced concrete.

Wall apertures

These weaknesses also occur for wall apertures. As masonry cannot accept tensile or bending loads, it is impossible to build across a wall aperture without support. Additional beams are needed to resist the loads and transfer them transversely into the adjacent parts of the wall. As steel does not meet fire protection criteria, these beams are usually made of reinforced concrete and, like the ceilings, must have additional thermal insulation or be built using U-shells. These special parts can either be made on the spot, e.g. at the same time as the concrete ceiling is cast, or delivered to the site as prefabricated lintels, reinforced in the factory.

Tie bars / ring beams

U-shells can also be used to create peripheral tie beams and ring beams. Other factors, such as wind forces, cause tensile forces in a building. These are transferred by the ceilings as sheets and cannot be absorbed by the walls alone. Peripheral tie beams can be made in the form of reinforced concrete beams or U-shells under the ceiling, or of appropriately reinforced ceiling strips. They transfer forces for all external and transverse walls. In the case of ceilings with no sheet action or with sliding supports (e.g. under flat roofs) the peripheral tie beams should run round the whole building as a continuous ring (ring beam).

> \\Hint:
> L- or U-shells are available from brick manufacturers as prefabricated parts. As the name suggests, the L-shell is L-shaped, to support the ceiling. U-shells are used above wall apertures and to create ring beams. The cavity is filled with concrete on site (see Fig. 31).

Single-leaf masonry with internal insulation

To improve the wall structure's thermal insulation properties, insulation or thermal insulation rendering can be applied to its internal side. This construction is problematical in terms of building science, however, as there is a danger that condensate will form on the inside of the cold masonry and impregnate the construction with moisture, which may lead to mould formation. For this reason, this method tends to be used for refurbishment, when it is not permissible to alter listed façades.

Laminated thermal insulation systems (LHIS)

To avoid these problems, the insulation is not fixed inside in a laminated thermal insulation system, but stuck onto the masonry and fixed with ties. To protect the insulation from the weather, however, a special layer of water- or moisture-resistant rendering is applied directly to the insulating material. As the rendering needs a solid ground, and as no holes or pressure points should be created by external factors, the insulation must resist compression and provide sufficient general resistance. LHIS is a common system for reasons of economy, above all when refurbishing existing buildings.

Single-leaf masonry with cladding

Another way of protecting loadbearing masonry is to suspend an outer skin in front of the building. This structure made of metal, wood or fibre cement can be attached directly to the masonry, or a space can be left for an additional insulating layer. Care should be taken with the fixing points, which could cool the masonry, and adequate rear ventilation to prevent moisture impregnation from water that gets behind the cladding.

Basement walls

Basement walls are single-leaf in all structures. An approach using waterproof reinforced concrete ("White Tub") is increasingly common, but another wall structure may be preferable. Basement walls need to be well reinforced against soil pressure, which affects the surface of the wall vertically, and against load transfer. When fixing dimensions, wall height, soil pressure and the superimposed load from the surface of the terrain should be taken into account. The thermal insulation that is applied outside (perimeter insulation) also has to be able to stand up to the soil pressure in working basement spaces with high thermal insulation demands, and must therefore be compression resistant. It can be made of sheet foamed glass, polystyrene particle foam or extruded polystyrene foam sheets. Basement walls must also be sealed against moisture in the soil. A concrete finish is preferred if there is heavy potential pressure from water, but if the load is less and the water does not exert pressure, horizontal and vertical sealing should be provided. The horizontal membrane in the form of a sealing sheet should be applied to the full area of the concrete slab, and must join up with vertical sealing in the form of

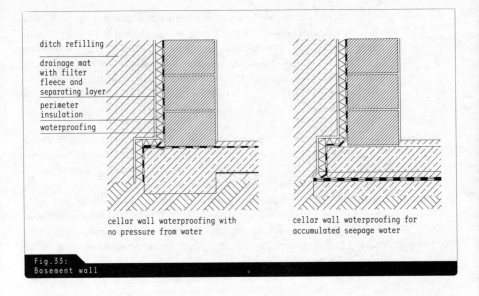

Fig.33: Basement wall

sheets or bituminous coatings on the on the outside of the wall under the first row of bricks. Finish as a "Black Tub" provides additional protection: here the horizontal membrane is attached onto a base course under the floor slab and given a protective coating. Both the vertical membrane and any possible thermal insulation can be protected against soil damage when the excavation pit is filled, by using a protective layer of geotextile membrane and filter fleece, which also drains off water.

Plinth zone

The plinth zone is more heavily loaded than the masonry above it by the adjacent soil and the effects of splash water. Hence, it should be sealed against moisture by a vertical membrane to a height of 30 cm above the top edge of the terrain. This ends with a horizontal damp course the full width of the wall, which prevents moisture from rising further into the masonry above it. This damp course should be protected by a row of weatherproof bricks, by cladding, or by applying a special water-resistant plinth rendering. The transition between the renderings can be carried out through the structure or by using differences in smoothness. Plaster bases, e.g. in expanded metal, help to avoid cracks at this point.

Double-leaf masonry

In double-leaf masonry, a second wall (external or facing leaf), which protects the inner side from the weather, is built in front of an inner wall (inner leaf), which has the primary loadbearing function. A gap is left

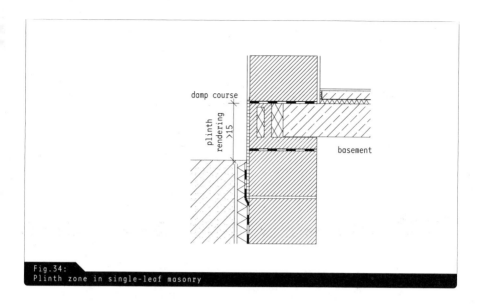

Fig.34:
Plinth zone in single-leaf masonry

Double-leaf masonry with cavity

between them (cavity), which can be left free, or wholly or partially filled with insulation.

The cavity is there to prevent water penetrating directly into the inner leaf and thus into the interior, and causing damage such as mould formation. If moisture has penetrated the outer leaf, it is removed via the cavity. To this end, ventilation apertures should be placed in the plinth area and at the top of the wall and wall apertures. These are usually open perpends with a horizontal damp course. This is achieved by laying a sealant strip or film as a "Z-barrier" across the full area of the bed joint below the open settlement joints, and taking it to the inner leaf with an incline of 1–2 cm and then 15 cm upwards.

\\ Important:
For all sealing membranes, special attention should be paid to points at which a wall or ceiling is penetrated by sanitation pipes or service connections, which must be carefully sealed.

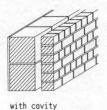 with cavity

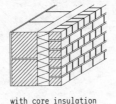 with core insulation

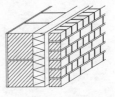 with insulation and cavity

Fig.35:
Double-leaf masonry

To ensure adequate back ventilation, the air gap should be at least 60 mm wide, or 40 mm if the joint mortar is cleaned off or if insulation is used. Although vertical air gaps – including the back ventilation – conduct little heat, for thermal insulation it is usually necessary to fit an insulating layer in the gap. If the entire gap between the leaves is filled, this is called double-leaf masonry with core insulation.

Full-fill cavity walls

This version increases the resistance to heat transfer, but not the thickness of the brick, and thus the thickness of the whole wall. The insulation can be in the form of blankets or strips fastened to the inner leaf, or loose granules or mixtures, which are shaken into the gap; care must be taken to distribute them evenly. The disadvantage of this structure lies in the fact that water can get in behind the front leaf. It is difficult to remove, and reduces the thermal insulation properties of the structure, as damp building materials transfer heat better than dry ones. The insulation material must therefore be permanently water-resistant, and joints and connection points must prevent water from penetrating. Softer mineral fibre strips should be packed tightly, and plastic foams given a stepped rebate or tongue and groove. Any damage caused by fixing the blankets or the outer leaf should be sealed. If insulation material is poured or shaken in, care should be taken that no material can fall out of the drainage apertures, e.g. by installing a rustproof perforated grille.

Double-leaf masonry with insulation and air gap combines the advantages of the two above-mentioned constructions.

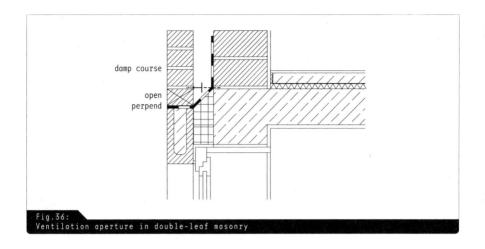

Fig.36:
Ventilation aperture in double-leaf masonry

Partial-fill cavity walls

A layer of water-resistant thermal-insulation blankets or mats is attached to the inner leaf and separated from the outer leaf by an air gap of at least 4 cm. This is more elaborate to build than other constructions. As loadbearing, insulation, damp and weather protection are strictly separated, it offers the best properties, but the whole structure will be thicker.

Inner leaf

In all constructions, the inner leaf serves mainly to provide structural stability and transfer load. It can be built with loadbearing bricks of a high specific density; these have low resistance to heat transfer, but offer a high level of sound insulation. Essentially, all standard bricks and mortars approved by the building authorities can be used for the inner leaf. › see chapter Building materials As the inner side usually has a layer of internal rendering applied to it that covers the bricks, large blocks can be used, running counter to the bonding rules, as they are built in random bond and with a thin mortar bed, but are very strong. Supporting concrete ceilings is not a problem for insulated versions. Thermal insulation can be placed continuously in front of the inner leaf. The full area of the ceiling can be supported by it and thermal insulation can be additionally improved where appropriate by placing an insulating strip in front of it.

External leaf

The external leaf protects the rest of the masonry from external factors and the weather. For this reason, only materials should be used that are appropriate for these conditions and are not sensitive to frost, moisture and the effects of being on the outside. Such units are offered by brick,

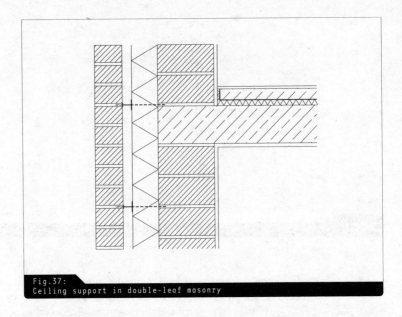

Fig.37:
Ceiling support in double-leaf masonry

calcium silicate and concrete block manufacturers as frost-proof, facing or vitrified units. ⟩ see chapter Building materials Mortar manufacturers also offer special frost-resistant mortars that absorb little water and are low in efflorescence, i.e. do not discolour as a result of salt deposits.

The outer leaf determines the appearance of the building and is ideally built in the commonly taught bonds described above. But this leaf can absorb only its own self-weight and has to be fixed to the inner leaf by wire anchors to secure it against wind pressure or suction, and avoid tipping over, collapsing or bulging. The number of anchors needed and their diameter depend on the distance between the leaves and the height of the wall. Separate attention should be paid to open edges of apertures, corners of the building or expansion joints, as well as rounded parts of the structure. The appropriate measures must be taken to prevent moisture from being transported from the outer to the inner leaf, such as fitting plastic discs so that the water can drip off in the gap.

Underpinning

In addition to wall-anchor fixing, the outer leaf must be regularly underpinned and attached to the inner leaf, so that at greater heights the self-weight can be transferred evenly into the loadbearing leaf, as well as being supported by the base. Rustproof bracket anchors and angle-brackets or thermally isolated ceiling projections are used for this.

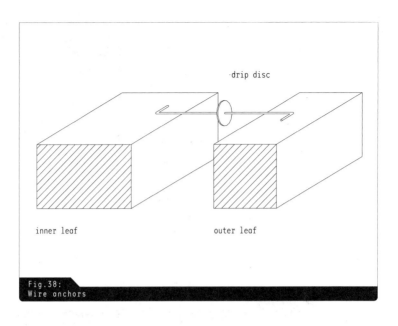

Fig.38:
Wire anchors

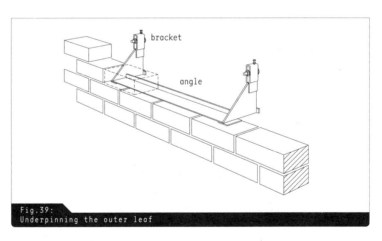

Fig.39:
Underpinning the outer leaf

The leaf must be secured against slipping away at the base. The first run of anchors should therefore be placed as low as possible. The lower sealing strip should extend to the front edge of the outer leaf.

The minimum thickness for the outer leaf is 9 cm. Anything thinner is referred to as wall cladding. › see chapter External walls For reasons of space

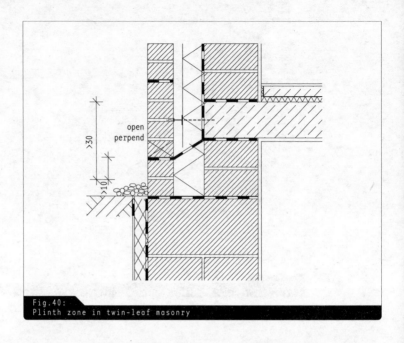

Fig.40:
Plinth zone in twin-leaf masonry

and thus of expense, the outer leaf is usually half a brick thick, so the visible bonds are not regular, as the most commonly taught bonds (except the stretcher bond) cannot be built in this way.

Apertures

For aesthetic reasons, the bond should generally run throughout the area of the wall. Apertures, windows and doors and any projections therefore need special anchors to hold the units in position. Lintels are often built in soldier bond, which is however not a regular construction, unlike the arch constructions described above, and cannot carry any load. The bricks should therefore be supported by brackets, which is cheap, but visible from the outside. Or there may be an invisible joint reinforcement to hold the bricks in place. Brick manufacturers also offer U-shells, which are reinforced and filled with concrete. These constructions transfer the load into the wall areas adjacent at the sides. All metal parts should be rustproof, ideally made of stainless steel, as galvanized items can be damaged in transport or fitting, and flaws are hard to see or reach after fitting.

Joints

The outer leaf deforms differently from the inner leaf as a result of temperature and weather. Vertical and horizontal movement joints should

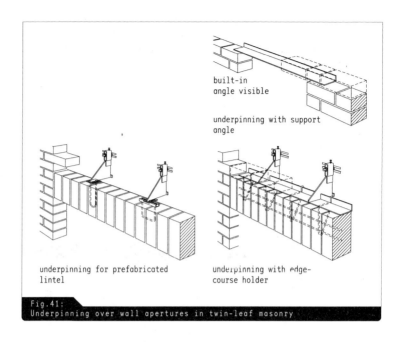

Fig. 41:
Underpinning over wall apertures in twin-leaf masonry

therefore be planned for the outer leaf to absorb this deformation. As well as the material-dependent distances between the expansion joints, › see Table 1 the walls should be separated at the corners on the basis of factors relating to the points of the compass. The west wall expands most, and the north wall least. These joints can however be offset by half the gap between the joints towards the middle, if this is architecturally desirable. Cracks around window sills caused by different loading of sill and the masonry around it can also be prevented by expansion joints on both sides. Structural reinforcement in the upper sill area may replace these joints. Horizontal joints should always be planned under the underpinning.

Non-loadbearing external walls – free-standing walls

Free-standing walls are very restricted in terms of height, as they are supported only at the base point and have no stabilizing imposed load. The walls must therefore be thicker or stabilized by crosswalls or columns. As they are outdoors and exposed to frost, they must use frost-resistant materials and foundations and be protected against moisture. Horizontal dampproof courses are needed above ground level, and the top of the wall should be protected by blocks, metal sheeting or concrete coping and damp courses.

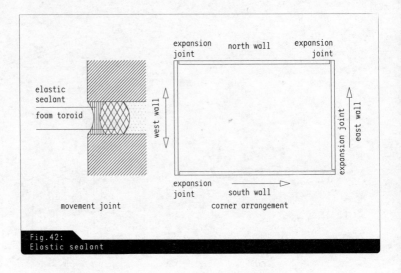

Fig.42:
Elastic sealant

Table 1:
Distance between joints

Masonry in	Distance between the expansion joints in m
calcium silicate brick, aerated concrete block, concrete block	6–8
lightweight concrete block	4–6
brick	10–20

From: P. Schubert: Zweischalige Aussenwände – Dehnungsfugen in der Aussenschale (Verblendschale), in: Mauerwerk 6/2003, Ernst & Sohn, Berlin, p.203

Figure 44 shows a comparison between different wall structures using the same brick. The lower the given thermal transfer coefficient U, the better the thermal insulation. The relationship of the results is more important than the precise value.

INTERNAL WALLS

Internal walls are not directly connected to the outdoors. They are already protected from cold, rain and snow by the external walls, ceilings and floors. Their main function is to separate internal areas, use zones or sightlines. The separation may require greater sound insulation, e.g. between dwellings, between bedrooms and living areas, between office and production areas, or it may have a fire protection function.

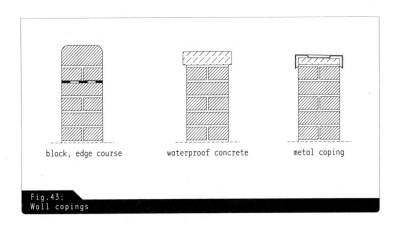

Fig.43:
Wall copings

Some inner walls also have to carry part of the load of the building, or stiffen the building or individual sections of wall. They can thus be loadbearing in direct connection with the adjacent structural elements, or non-loadbearing, in which case all they have to do to avoid falling over is transfer their self-weight and the horizontal loads on their area to other structural elements. These different requirements are reflected in both the dimensioning and the detail of the connection points. Specific density affects compressive strength and above all sound insulation, and plays a key part in relation to internal walls. Here, units with a high mass and specific density offer both great compressive strength and good sound insulation.

Loadbearing and stiffening internal walls

Loadbearing internal walls stiffen the building and provide ceiling supports. To stiffen a wall, the connection with it should be tension- and compression-resistant; building materials with approximately the same deformation behaviour should be chosen where possible for the sake of stability. Connection is achieved by building both walls to the same height in bond or by leaving gaps (socket connection) or protruding bricks (projection connection) in the wall to be stiffened, which will be worked on

\\ Hint:
The thermal transfer coefficient is the inverse value of the sum of all thermal transfer and transition coefficients.

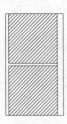
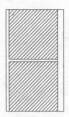
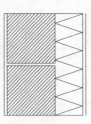

with external
rendering

light mineral rendering
2 cm 0.31 W/mK
transverse brick 30 cm 0.14 W/mK
internal rendering 1.5 cm 0.7 W/mK

U=0.417 W/m²K

with thermal insulation
rendering

thermal insulation rendering
3 cm 0.07 W/mK
transverse brick 30 cm 0.14 W/mK
internal rendering 1.5 cm 0.7 W/mK

U=0.362 W/m²K

compound thermal
insulation system

compound thermal insulation
system 6 cm 0.035 W/mK
transverse brick 30 cm 0.14 W/mK
internal rendering 1.5 cm 0.7 W/mK

U=0.234 W/m²K

concrete

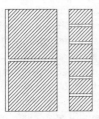
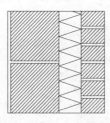
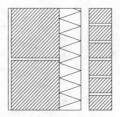

with cavity

facing 11.5 cm 0.68 W/mK
cavity, wire anchors 4 cm
transverse brick 30 cm 0.14 W/mK
internal rendering 1.5 cm 0.7 W/mK

U=0.412 W/m²K

with core insulation

facing 11.5 cm 0.68 W/mK
core insulation 6 cm 0.35 W/mK
transverse brick 30 cm 0.14 W/mK
internal rendering 1.5 cm 0.7 W/mK

U=0.236 W/m²K

with insulation and cavity

facing 11.5 cm 0.68 W/mK
cavity (wire anchors) 4 cm
core insulation 6 cm 0.035 W/mK
transverse brick 30 cm 0.14 W/mK
internal rendering 1.5 cm 0.7 W/mK

U=0.242 W/m²K

Fig.44:
Comparison of wall construction types

subsequently. The stiffening wall can thus be erected later, an advantage if additional space is needed, e.g. for scaffolding. However, this method does require additional reinforcing bars in the joints to absorb the tensile forces.

An efficient alternative is butt walling, which also requires tensile bars or anchors; the joint is pointed subsequently. This connection can be used only at internal corners, and has the advantage that when connecting with the external walls the thermal insulation of the external wall is not compromised by the intrusion of interlocking bricks from the internal wall, which could be made of different materials.

Party walls

Walls separating adjacent dwellings must always be twin-leaf structures for sound insulation purposes. The cavity width depends on the mass of the partitioning leaves; a width of 5 cm is recommended. The cavity should be filled with tightly packed mineral fibre blankets covering the full surface. Sound insulation is improved further by lagging in two layers with offset seams. Rigid foam sheets are inadmissible. Care should always be taken that no mortar drops into the joint. When building or attaching ceilings, the insulation should always be continued above the edge of the wall or ceiling.

Non-loadbearing internal walls

Non-loadbearing internal walls may not be used for either stiffening or load transfer, and must not be subjected to wind loads. They carry all their self-weight and light bracket loads (e.g. shelves, pictures etc.), and must transfer impact loads to adjacent structural elements. Wall lengths have been calculated according to height, the way in which the wall is attached to adjacent structural elements (two-sided to four-sided mounting), and possible imposed loads from ceiling deformations. They are presented in a table that may be used without acknowledgement.

Connections to adjacent structural elements can be rigid or sliding. Rigid connections should be used when there is little load from other structural elements that could lead to indirect stresses. They have good sound insulation and fire protection properties, and are inconspicuous as they are carried out without mortaring, steel inlays or interlocking.

Sliding connections are made using steel sections or sliding joint anchors and can absorb some deformations. These connections are very elaborate and may be visible, or need to be covered.

> \\ Important:
> Special calculations must be applied to masonry with unpointed perpends!

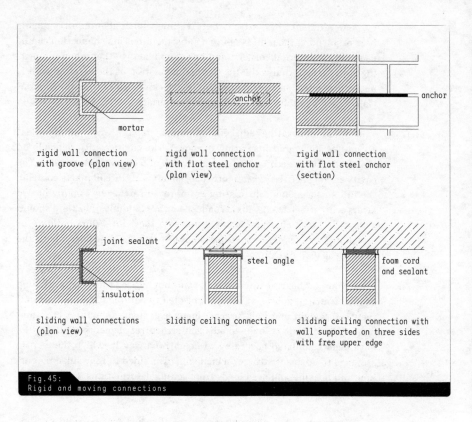

Fig.45:
Rigid and moving connections

SLOTS AND GAPS

When dimensioning walls, care should be taken when compromising the cross-section with slots and holes, e.g. for electrical or sanitary installations. Limiting values must not be exceeded. Many manufacturers offer special masonry units that already provide apertures for installations.

BUILDING MATERIALS

Earlier chapters describe masonry units in general terms and distinguish them only in terms of dimensions, geometry or the place where they are to be used. The question now arises of which material is suitable for a particular construction and the functions it has to perform. Various kinds of brick and mortar are listed and distinguished below.

MASONRY UNIT TYPES

In addition to the many types of natural stone, which will not be considered in detail here because they are so seldom used as pure masonry units in building construction, there is an equally wide variety of artificially manufactured bricks and blocks. To fulfil the functions of masonry – supporting, separating, facing, insulating, protecting – they are finished in a variety of ways, and have many different properties. The following summary can be made on the basis of the above-mentioned relationship between the unit's specific density and the requirements:

high dry specific density	=	good compressive strength
high dry specific density	=	good sound insulation
low dry specific density	=	good thermal insulation

Masonry standards: harmonized European product standard

The harmonized European product number series EN 771 (Specification for masonry units) is valid in the EU states. The series consists of:

EN 771-1 Clay masonry units
EN 771-2 Calcium silicate masonry units
EN 771-3 Aggregate concrete masonry units
EN 771-4 Autoclaved aerated concrete masonry units
EN 771-5 Manufactured stone masonry units (dense and lightweight masonry units)
EN 771-6 Natural stone masonry units

The standards establish basic specifications for source materials, manufacture, requirements, description and testing of masonry units. They do not fix precise sizes, nominal dimensions and angles. To be traded in Europe, construction products in these categories must carry the CE mark as a sign of compliance with the standards. Approval of products, and thus permission to use them, is still a national responsibility.

> 📎
\\Hint:

In Germany, the so-called user standards are used to translate the CE classification values to make them compatible with national standards. These provide precise values or admissible fields for the declared specifications (DIN V 20000-401 to DIN V 20000-404). Since the European standards do not address some requirements that have already been introduced, which are therefore not covered by the usage standards, so-called residual standards were added for the sake of completeness:

DIN V 105-100   Clay units with specific properties
DIN V 106-100   Calcium silicate units with specific properties
DIN V 4165-100 Autoclaved aerated concrete units – high-precision units and elements with specific properties
DIN V 18151-100 Lightweight concrete hollow blocks – hollow blocks with specific properties
DIN V 18152-100 Lightweight concrete solid bricks and blocks – solid bricks and blocks with specific properties
DIN V 18153-100 Concrete masonry units – masonry units with specific properties

These also regulate all previously valid product properties, characteristics and differentiations and lay down precise values and specifications in table form, e.g. for compressive strength and specific density classes, and for unit perforations.

> 📎
>
> Clay masonry units

STANDARD MASONRY UNITS

The clay brick is one of the oldest artificial building materials in the world. Bricks were made as long as 4000 years ago in the Haruppa cities on the Indus, and even then they had roughly the same dimensions and shape as today's standard brick. At first, mud bricks were baked in the sun, then the fired clay brick developed into a high-tech product, porous and thus offering excellent heat insulation when made with combustible aggregates, or protection against the elements when fired to the point of sintering; it gave us our current image of masonry. Its form, finish and material were developed even further, and now come in a wide variety of units, with form and performance fixed precisely by standards. The brick stands for both a long tradition of craftsmanship and a progressive and economical building material. It is made by mixing loam and clay, pressed and extruded as a ribbon, cut into appropriate sizes and fired.

The harmonized standard EN-771 makes a distinction between LD and HD bricks, and divides them into categories I and II, which fix a tolerance limit for maintaining compressive strength and thus quality. To be classified in category I the probability of deviating from the declared compressive strength must not be above 5%. All the rest of the units in category II are no longer accepted by the national standards.

Fig.46:
Clay masonry units

LD bricks are used mainly by the internal loadbearing leaf of a twin-leaf structure or for rendered single-leaf masonry, as they have a low dry density (<1000 kg/m^3) and thus good thermal insulation properties. This is achieved by adding polystyrene beads or sawdust that burn when the brick is fired and leave tiny pores. They may be used only for masonry protected from penetrating water.

HD bricks with a gross dry density of >1000 kg/m^3 are suitable for both protected and unprotected masonry. This includes resistant units for the outer leaf and heavy sound-insulating units for the internal walls.

In these categories we distinguish:

Solid bricks	HD bricks with perpendicular perforation that takes up a maximum of 15% of the bed face or 20% of the volume.
Vertically perforated bricks	LD or HD bricks with vertical perforation of between 15% and 50% of the bed face. Here a distinction is made between perforation types A, B, C and W.
Heat insulation bricks	LD bricks with higher thermal insulation specifications and a special perforation type.
Solid vertically perforated facing bricks	A category of brick that is frost-resistant as well as meeting the above perforation specifications.
Solid and vertically perforated engineering bricks	HD bricks with a vitrified surface. They absorb only minimal quantities of water, have a compressive strength of at least class 28, are frost-resistant and have higher specific density requirements. Here a distinction is made according to the above-mentioned criteria between solid units and vertically perforated

Table 2:
Bricks

Material:	Clay, loam, clayey masses		
Aggregates:	Sawdust, polystyrene beads (optional)		
Manufacture:	Moulded and fired		
Dimensions	In mm (e.g. 240 x 300 x 238) and in multiples of DF (e.g. 10DF)		
Unit types		Strength class*	Density class*
LD bricks	Vertically perforated brick	6–12	0.7–0.9
	Thermal insulation brick		
HD bricks	Solid brick	8–28 (36**)	1.6–2.2
	Vertically perforated brick	8–20 (36**)	1.2–1.6
	Solid facing brick	8–28 (36**)	1.8–2.2
	Vertically perforated facing brick	8–28 (36**)	1.2–1.6
	Solid engineering brick	28	1.8–2.2
	Vertically perforated engineering brick	28	1.8–2.2
	Solid engineering brick	60	1.8–2.2
	Vertically perforated high-strength engineering brick	60	1.8–2.2
	Panel brick		

* Common classes
** Values for high-strength bricks or engineering bricks (without special abbreviations)

units with holes A, B. C. High-strength engineering bricks must achieve a compressive strength of at least class 36.

High-strength engineering bricks — These have a compressive strength of at least class 60 and a specific density of 1.4. They are particularly resistant and durable.

Panel bricks — These have channels to take mortar or concrete when constructing reinforced masonry.

Additional stipulations concern the shape of grip openings that make the bricks easier to handle, or the form of mortar pockets or tongue and groove systems that work without visible mortar application to the perpends.

Calcium silicate units

Calcium silicate units have been made only since they were patented in 1880. Unlike bricks, they are not fired. Instead, a mixture of sand, water and lime is hardened under high pressure.

Fig.47:
Calcium silicate units

Table 3:
Calcium silicate units

Material:	Lime, sand (quartz sand), water		
Aggregates:	Dyes and additives		
Manufacture:	Mixed, moulded and hardened under pressure		
Dimensions	In mm (e.g. 240 x 300 x 238) and in multiples of DF (e.g. 10DF)		
Unit types		Strength class*	Density class*
	calcium silicate solid brick	12–28	1.6–2.0
	calcium silicate perforated / hollow block	12–20	1.2–1.6
	calcium silicate facing brick, solid brick	12–28	1.6–2.0
	calcium silicate facing brick, solid brick	20–28	1.6–2.0
	calcium silicate facing brick / perforated brick	12–20	1.4–1.6
	calcium silicate facing brick, perforated brick	20	1.4–1.6
	calcium silicate prefabricated bricks		
	calcium silicate prefabricated elements		

* Common classes

As for clay bricks, a distinction is made between solid calcium silicate bricks and perforated calcium bricks according to the proportion of holes: the upper limit is 15% of the bed face. Both sorts must have a unit height of less then 113 mm. Higher units are called calcium silicate blocks or hollow calcium silicate blocks. Calcium silicate facing bricks and calcium silicate engineering bricks are available for masonry exposed to weathering. Calcium silicate prefabricated bricks are available where appropriate for laying in thin-bed mortar and calcium silicate R units, which require no mortar for their perpends because of their tongue and groove system.

› see chapters Rules of construction, Finishing rules, and New approaches

Fig. 48:
Porous concrete units

Table 4:
Aerated and light concrete units

Material:	Lime, quartz sand, cement, water, expanding agent to form pores (aluminium)		
Aggregates:			
Manufacture:	Mixed, moulded and hardened under pressure		
Dimensions	In mm (e.g. 240 x 300 x 238)		
Unit types		Strength class*	Density class*
	Aerated concrete block unit	2-4	0.4-0.7
	Aerated concrete prefabricated unit	2-4	0.4-0.7
	Aerated concrete slab	Non-loadbearing	
	Prefabricated aerated concrete slab	Non-loadbearing	

* Common classes

Aerated concrete units

This type was also developed in the late 19th century. For the manufacture of aerated concrete units, a mixture of quartz sand, lime and cement is poured moulds with water and provided with steel mesh reinforcement according to purpose. Powdered aluminium is used as an expanding agent, increasing the proportion of pores to 90% of the material's volume through the release of hydrogen. The unmoulded material is cut and hardened under pressure.

Porous concrete is like the natural mineral tobermorite and is offers high thermal and sound insulation because of its high porosity.

Large-format aerated concrete blocks or prefabricated units for thin-bed mortar are used for loadbearing walls.

Aerated concrete slabs and prefabricated slabs are used only for non-loadbearing walls with different loadbearing systems, and for sound insulation walls.

Fig.49:
Concrete and lightweight concrete units

Table 5:
Concrete and lightweight concrete bricks and slabs

Material:	Mineral aggregates and hydraulic binding agents			
Aggregates:	Pumice, expanded clay for lightweight concrete			
Manufacture:	Mixed, moulded			
Dimensions	In mm (e.g. 240 x 300 x 238) and in multiples of DF (e.g. 10DF)			
Brick types	Category		Strength class*	Density class*
	Concrete bricks			
		Solid concrete bricks	12-20	1.6-2.0
		Solid concrete bricks	12-20	1.6-2.0
		Hollow concrete bricks	2-12	0.8-1.4
		Concrete facing bricks	12-20	1.6-2.0
		Concrete facing block	12-20	1.6-2.0
	Lightweight concrete bricks			
		Solid lightweight concrete bricks	2-6	0.6-2.0
		Solid lightweight concrete blocks	12	1.6-2.0
		with slots		
		with slots and special thermal insulation properties		0.5-0.7**
		Prefabricated bricks		
		Hollow	2-6	0.5-0.7
		Lightweight concrete wall elements	Non-loadbearing	
		Hollow lightweight concrete wall elements	Non-loadbearing	

* Common classes
** Standard specification – to clarify special thermal insulation properties

Storey-height elements and ceiling slabs complete the product programme as additions to classical masonry construction.

Concrete and lightweight concrete units

Concrete and lightweight concrete units are cast in moulds and stored until they reach their ultimate strength. The difference between the two lies in the nature of the aggregates. Only lightweight aggregates with a porous microstructure (primarily natural pumice or expanded clay) may be used for lightweight concrete.

A distinction is made here in terms of dimensions as well as aggregates. Solid bricks are limited to a height of 115 mm, which distinguishes them from solid blocks, which are 175 mm or 238 mm high. Neither type is permitted to have cells, but only grip openings. Hollow blocks with a preferred height of 238 mm do have cells, whose number precedes the unit category (e.g. 3K). Facing units or facing blocks must be used in situations with weathering.

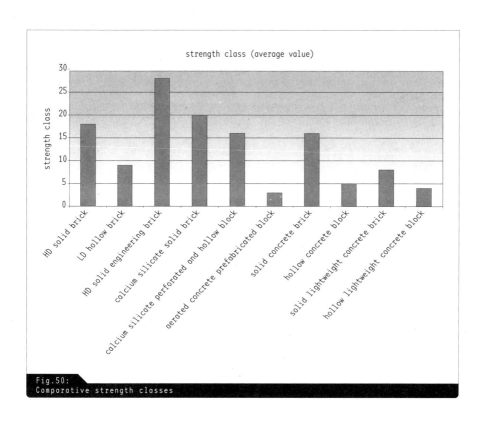

Fig.50:
Comparative strength classes

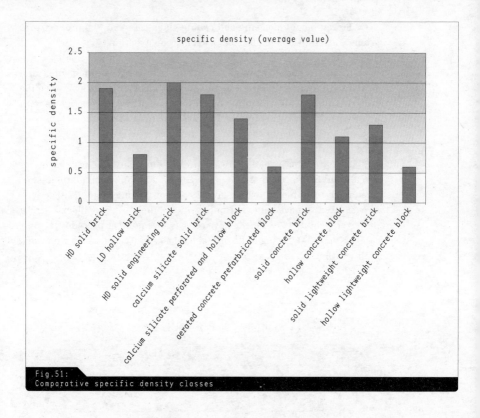

Fig.51:
Comparative specific density classes

Lightweight concrete units are distinguished according to the same criteria between solid bricks, solid blocks and hollow blocks. There are also bricks with slots and special insulating properties, identified by the endings -S or -SW, and prefabricated bricks.

Lightweight concrete wall construction elements and hollow wall elements are manufactured for non-loadbearing walls.

\\Important:
Site-mixed mortar is not covered by European standards. National user or working standards must be consulted here.

Table 6:
Mortar types

Mortar type Abbreviations according to EN 998-2	Mortar class according to EN 998-2 (only CE sign)	Forms available
Normal-weight mortar (G)		Premixed dry mortar Ready-mixed mortar Multi-chamber silo mortar (building site mortars in Germany)
	M2,5	
	M5	
	M10	
	M15	
	M30	
Lightweight mortar L		Premixed dry mortar Ready-mixed mortar Multi-chamber silo mortar
	M10	
	M10	
Thin-bed mortar T		Premixed dry mortar
	M15	

TYPES OF MORTAR FOR MASONRY

Mortar is made up of binding agents, admixtures and additives. Admixtures affect mortar properties such as frost resistance or workability, and may be added in larger quantities. Additives change the properties of the mortar through chemical and physical processes and may be used to a limited extent only. They include liquefiers, retarders and air entrainers. The components are supplied either individually and mixed on site (site-mixed mortar), or are delivered to the site ready-mixed.

All the components except the water can be supplied ready-mixed (premixed dry mortar), or to save time the ready-made mortar can be supplied to the site from the factory. Retarders allow for the necessary working time (ready-mixed mortar). For premixed dry mortar, only the non-hardening materials are mixed, so water and cement have to be added on site. One variant of ready-mixed mortar is supplied as multi-chamber silo mortar. Here the components are mixed on site as well, but without the possibility of altering the mixing ratio.

\\ Important:
Similarly to bricks, the national standards DIN V 20000-412 (user standard) and DIN 18580 (residual standard) apply additionally in Germany. But essentially the specifications of DIN 1053-1 still apply.

\\ Important:
Masonry mortars show considerable discrepancies between the European standard and the stipulations of DIN 1053 in Germany. Here precise attention must be paid to the user and residual standards!

Just like masonry units, mortar for masonry is subject to precisely specified manufacture, inspection, categorization and property definition. The harmonized product standard DIN EN 998-2 applies in the EU.

This divides masonry mortar into three types: normal-weight mortar (G), lightweight mortar (L) and thin-bed mortar (T).

Normal-weight mortar differs from lightweight mortar in terms of its dry gross density m, which must be at least 1500 kg/m^3, while lightweight mortar has a dry density of less than 1300 kg/m^3. Thin-bed mortar was developed specially for gauged bricks and reduces the mortar height to 1–3 mm. Here the dry density may not be less than 1500 kg/m^3 and the maximum aggregate particle size is 2 mm. All mortar types are allocated to the mortar groups M1–M30 according to their compressive strength; the compressive strength value is given in N/mm^2. If a mortar complies with DIN EN 998-2 it is marked with the CE sign.

IN CONCLUSION

The information contained in this book can give only a rough guide to the many possibilities offered by masonry construction. For this reason, it does not generally cover the regulations laid down in the different national standards, which sometimes differ. These will have to be addressed separately, using the list of standards in the Appendix. But the knowledge presented does provide the necessary basis for understanding the essential rules of the craft and the areas in which masonry is used, and makes it possible for the reader to continue independently.

Many regular constructions can be explored using the principles shown in the second chapter, which present an extensive design repertoire for the planner. The constructions listed in the third will make it easier to put the legal rulings and standards in context. Problems arising from related topics such concrete construction or façades, or more advanced expositions of structural behaviour or building science, will be more readily understood. Information from manufacturers and dealers, which the internet is turning into an increasingly wide and important reference source for planners, is made easier to filter by the details given in the fourth chapter, when making selections for future building commissions.

All in all, these "Basics" make it possible to explore the diverse world of masonry construction and approach it correctly.

APPENDIX

STANDARDS

Masonry units:

EN 771-1 (consult national versions)	Specifications for bricks – Part 1: Clay masonry units
EN 771-2 (consult national versions)	Specifications for bricks – Part 2: Calcium silicate masonry units
EN 771-3 (consult national versions)	Specifications for bricks – Part 3: Aggregate concrete masonry units (dense and lightweight aggregates)
EN 771-4 (consult national versions)	Specifications for bricks – Part 4: Autoclaved aerated concrete masonry units
EN 771-5 (consult national versions)	Specifications for bricks – Part 5: Manufactured stone masonry units
EN 771-6 (consult national versions)	Specifications for bricks – Part 6: Natural stone masonry units

Masonry mortar:

EN 998-2 (consult national versions)	Specifications for mortar in masonry structures – Part 2: Masonry mortar

Other building parts and materials:

EN 845-1 (consult national versions)	Specifications for additional parts for masonry – Part 1: Anchors, tie members, bearings and brackets

Loads and forces:

EN V 1996-1-1	Eurocode 6: Dimensioning and constructing masonry buildings Part 1-1: General rules – rules for reinforced and non-reinforced masonry

LITERATURE

Andrea Deplazes (ed.): *Constructing Architecture*, Birkhäuser Publishers, Basel 2005

Francis D.K. Ching: *Building Construction illustrated*, 3rd edition, John Wiley & Sons, 2004

Ernst Neufert, Peter Neufert: *Architects' Data*, 3rd edition, Blackwell Science, UK USA Australia 2004

Andrew Watts: *Modern Construction Roofs*, Springer, Wien New York 2006

Günter Pfeifer, Rolf Ramcke, Joachim Achtziger, Konrad Zilch: *Masonry Construction Manual*, Birkhäuser Publishers, Basel 2001

Jacques Heyman: *The Stone Skeleton: Stuctural Engineering of Masonry Architecture*, Cambridge University Press, Cambridge 1995

Theodor Hugues, Klaus Greilich, Christine Peter: *Detail Practice: Building with Large Clay Blocks and Panels*, Birkhäuser Publishers, Basel 2005

Construction Products Directive: Directive of the Council of 21 December 1988 (89/106/EEC)

Kenneth Burke: *Perspectives by Incongruity*, Indiana University Press, Bloomington 1964

Andrea Palladio: *I Quattro Libri dell' Architettura*, English translation by Robert Tavernor, MIT Press, Cambridge, Massachusetts 1997

PICTURE CREDITS

Illustration page 10: Bert Bielefeld, Nils Kummer
Illustration page 34: Bert Bielfeld, Nils Kummer
Illustration page 55: Gesellschaft Weltkulturgut Hansestadt Lübeck
Willy-Brandt-Allee 19
23554 Lübeck

Figures 1–51:	Nils Kummer
Figure 26:	supported by:
	Bert Bielefeld and Kalksandstein-Info GmbH (see Fig. 47)
Figures 39, 41	supported by:
	Deutsche Kahneisen GmbH
	Nobelstrasse 51-55
	12057 Berlin
	www.jordahl.de
Figure 44, 46:	supported by:
	Wienerberger Ziegelindustrie GmbH
	Oldenburger Allee 36
	30659 Hanover
	www.wienerberger.de
Figure 47:	supported by:
	Kalksandstein-Info GmbH
	Entenfangweg 15
	30419 Hanover
	www.kalksandstein.de
Figure 48:	supported by:
	Bundesverband Porenbetonindustrie e.V.
	Dostojewskistrasse 10
	65187 Wiesbaden
	www.bv-porenbeton.de
Figure 49:	supported by:
	Meier Betonwerk GmbH
	Industriestrasse 3
	09236 Claussnitz/OT Diethensdorf
	www.meier-mauersteine.de

导言

砌体结构具有如下独特的传统、风格和特点：形式灵活多样，既适用于古典建筑也适用于现代前卫建筑；风格变化与时俱进，既能满足普通建筑也能保证实验建筑的需要。尽管可以不采用传统的砌体结构作为现代高层建筑和现代玻璃建筑的结构方式，但实际上很难找到内部没有砌筑墙体的建筑，因而我们不得不承认，砌体结构在建筑中几乎无所不在。

本书将介绍有关砌体的基础知识及一些相关知识。首先，我们需要了解构成砌体的基本元素：砖和砂浆；第二章，基于理论和实践介绍了砌块和砂浆砌筑成墙体的标准和原则；第三章，对砌体结构进行了介绍和解释；从砌块到墙体，从而很自然地引出第四章的内容，适用于砌体的建筑材料，故最终又以砌块结束。

砌体

砌体作为一种建筑材料，与木材和钢材不同。它是由两种材料（砖和砂浆）按照一定的工艺标准施工处理后形成的复合材料。砌体作为一种复合材料，其性能更类似于钢筋混凝土而非混凝土，因而其质量依赖于两种组成材料的质量和施工操作的实际质量。

砌体主要用来建造墙体，作为承重墙或剪力墙，围护墙体或者梁和柱之间的填充墙体。砖也见于拱形结构、屋盖中，还可作为楼面材料。

砖和砂浆种类繁多，因而了解这两种材料的重要性能及其施工特点是极其重要的，这样我们才能获得理想的复合材料——砌体。

砖

一谈到砖，许多人就会绘制出形状和尺寸均匀一致的草图：标准砖。经过近千年的发展，砖特性影响了砌体的立面形式，并且与我们的砌体观念密切相关。砖有多种不同的形状和尺寸：罗马平砖、大尺寸机加工砌块或八边形模制砖。现在砖特性形成了建造砌体建筑的整套工艺标准基础。这些标准决定了房间的大小和整个建筑的尺寸，砌体缝隙效果和特点，以及结构立面形式。

砌筑砂浆

砌体的第二种构成材料是砂浆。砂浆涂覆在砖的表面，能够平衡

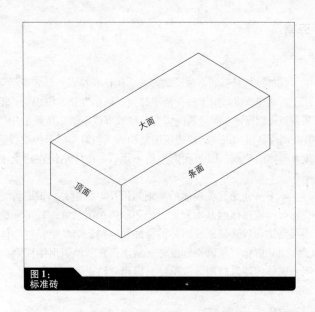

图1：
标准砖

砖块的尺寸公差，确保砌块牢固地粘结在一起，同时其抹平效果和颜色也决定了砌体的外观。水平方向上，砂浆应用于各层砖之间；竖方向上，砂浆施加于不同的砖之间。尽管由于成本和工程技术原因，现代砌体制作方法减小了各层之间的砂浆用量，但在规划施工时，砖和砂浆两者之间的粘结却是至关重要的。砖和砂浆的选择以及这两种材料之间的粘结效果对砌体承载力的影响十分显著。对采用现代建筑技术（砂浆用量很少或不含砂浆）的砌体亦是如此。

施工工艺标准

砌筑作为一种工艺，为了提高其质量，需满足一系列的标准。这样做的主要目的如下：
——提高砌体结构的承载能力和性能；
——减少材料损失；
——加快施工进度；
——提高设计质量，确保设计能充分发挥材料的性能和用途。
这些标准形成了墙体结构的理论基础（见"砌体结构"），揭示了砌体从组成材料到产生的原则和方法、合理尺寸，以及如何正确处理节点的洞口等。下面我们首先考虑单片墙体。

P11
尺寸和模数

建筑师在规划和创建一座建筑时的主要工作之一就是调整和协调各种结构和建筑工艺之间的关系。为了提高施工效率，骨架结构（墙、柱和楼盖等）和围护结构（门、窗、墙体和楼面料等）应互相匹配。重复利用相同尺寸的构件可使实际的建造过程同规划设计和装饰过程一样得到简化。由于要考虑砌块之间的砂浆灰缝，以及砌块尺寸的影响，固定网格尺寸会给砌体带来麻烦。这里采用一个简单的方法来确定墙体长度是否需要考虑砂浆灰缝的影响：专用尺寸和名义尺寸的区别。

专用尺寸和名义尺寸　　专用尺寸是一种基本的理论测量尺寸，网格和模数共同构成了整个砌体结构体系；名义尺寸是实际使用的尺寸，并出现于施工图中。我们采用上述两种尺寸（专有尺寸和名义尺寸）对灰缝的施工类型进行系统化，特别是对砌体灰缝施工进行系统化。

建筑没有灰缝时，其名义尺寸和专用尺寸完全相同，我们需根据施工中灰缝的形式进行如下考虑：

专用尺寸由名义尺寸和相应的灰缝构成，即砌块尺寸＋灰缝；
名义尺寸只包括砖尺寸，不包括灰缝尺寸。

现在设想一个带窗洞和横墙的砌筑墙体，我们马上会意识到由于砂浆灰缝导致墙体宽度、洞口尺寸和外伸段具有不同的名义尺寸和专用尺寸。

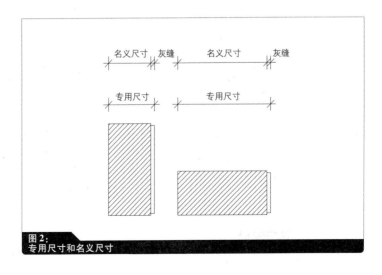

图2：
专用尺寸和名义尺寸

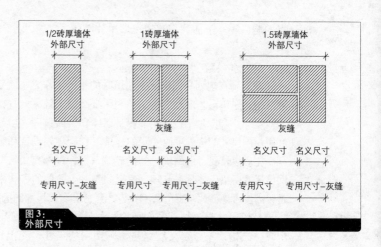

图3:
外部尺寸

外部尺寸　　　外部尺寸即为墙体厚度。不管砖的数量是多少,当减少一个灰缝时,专用尺寸都要减去一个灰缝的尺寸。

外部尺寸（E）＝专用尺寸－灰缝尺寸

洞口尺寸　　　一个洞口的内部尺寸通常包括一个附加的灰缝。

洞口尺寸（A）＝专用尺寸＋灰缝尺寸

凸出尺寸　　　凸出尺寸量测的是洞口和墙体之间以及洞口和外伸墙之间墙体段的长度。外部尺寸中减少的灰缝和洞口尺寸中附加的灰缝恰好相互抵消。

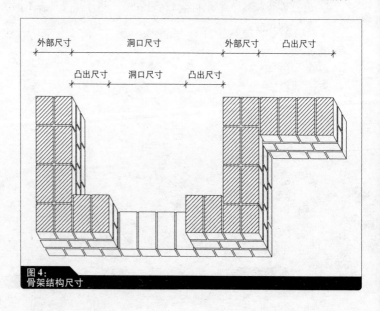

图4:
骨架结构尺寸

74

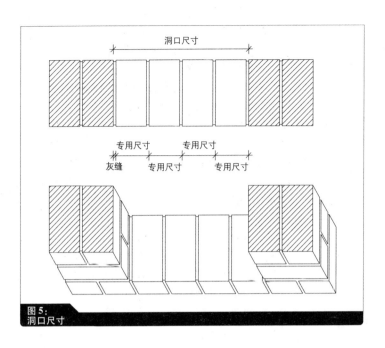

图5:
洞口尺寸

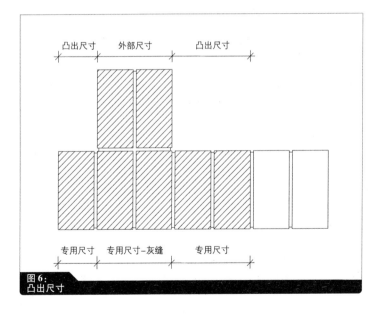

图6:
凸出尺寸

凸出尺寸（A）=专用尺寸

计量尺寸及名称

到目前为止，理论定义上仍然没有解决好实际尺寸问题，因为这些问题独立于砌块和灰缝尺寸而存在。这些尺寸根据地方惯例的不同而变化，因而在不同的国家形成了不同的标准。

在德国，砌筑工艺几乎无一例外地完全采用1/8m制作为专用尺寸（1m的1/8等于12.5cm）。标准砖，即普通规格的砖，尺寸为24cm×11cm×7.1cm（名义尺寸）。当竖向灰缝尺寸采用1cm而水平层间灰缝（简称横缝）尺寸采用1.23cm时，此时其专用尺寸为25cm×12.5cm×8.33cm，几层后尺寸达到1m。

砌体的灰缝尺寸在不断变化。新的加工制作工艺以及对砌体保温隔热、隔声和承载力的更高需要，导致砌筑工艺不再采用厘米级的砌筑灰缝。由于采用现代技术加工制作的砌块尺寸偏差很小，故采用几毫米厚的砌筑灰缝即可满足要求。

然而，为了保持常见的专用尺寸，需要调整单位尺寸以确保所有尺寸符合如下要求：

例如：

传统技术：	德国标准规格砖	24cm+1cm灰缝=25cm
现代技术：	加工砌块	24.7cm+0.3cm灰缝=25cm

常见的小尺寸砖规格如下：
长×宽×厚=24cm×11.5cm×7.1cm——普通规格（NF）

注释：
在德国建筑业，上述尺寸由DIN 4172尺寸标准确定。该标准从二战后的重建起，基于传统规格，对于骨架结构规定了一个25cm的基本模数。后来的DIN 18000建筑模数标准，采用更为简单的分米模数制-10cm，然而却没有流行起来。

提示：
不同国家基于本民族的传统习惯或不同的计量单位（例如英寸）具有不同规格的标准砖，例如：英国标准砖的尺寸为：21.5cm×10.25cm×6.5cm，比利时标准砖的尺寸为：19cm×9cm×6.5cm，美国标准砖的尺寸为：8英寸×4英寸×2.25英寸（20.3cm×10.2cm×5.7cm）。

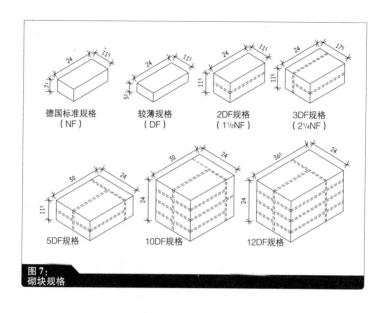

图7：砌块规格

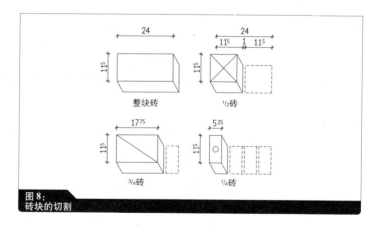

图8：砖块的切割

24cm×11.5cm×5.2cm—较薄规格（DF）

较大尺寸砖单元由较薄规格的砖和砂浆灰缝按照基本的模数砌筑而成，例如：5DF 由 5 块较薄规格砖经砂浆灰缝砌筑而成。

1/8 米制中包括的尺寸如下：

专用尺寸：12.5cm、25cm、37.5cm、50cm……100cm 等；

> **提示：**
> 由于同样数量的较薄规格砖可以不同方式组合，这样不同规格的砖能形成相同的尺寸，例如：
> 8DF＝24cm×24cm×23.8cm 和 8DF＝24cm×49cm×11.3cm

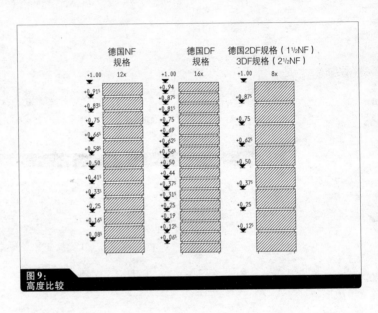

图9：
高度比较

名义尺寸：11.5cm、24cm、36.5cm、49cm……99cm 等；
外部尺寸：11.5cm、24cm、36.5cm 等；
洞口尺寸：51cm、1.01m、1.26m 等；
凸出尺寸：12.5cm、25cm、1.00m 等。
切割砖块时，一定要减去灰缝尺寸：
3/4 砖 ＝ 专用尺寸/4 × 3 − 灰缝尺寸 ＝ 6.25cm × 3 − 1cm
＝ 17.75cm。

砖块的切割详见砖块平面布置图所示，3/4 砖块（17.75cm）由对角线控制；1/2 砖块（11.5cm）由交叉的对角线控制；1/4 砖块（5.25cm）由一点或一圆控制。

对于砌体高度，通常采用 1/8m 制数值。为了获得砌体专用尺寸的高度（25cm、50cm 和 1m 等），水平砂浆灰缝（横缝）作为水准高程层，其尺寸大小介于1.05cm与1.22cm之间。

P18　　　　　**砖块组砌方式**

在砌体结构中，砖块的层称为皮。根据砖的铺设方向区别如下：

顺砖　砖块长边平行于墙体轴线方向；

丁砖　砖块长边垂直于墙体轴线方向；

边砖　砖块横向砌筑，长边朝下站立；

立砖　竖向丁砖砌筑，砌块窄边朝下站立。

不同情况下，当采用顺砖和丁砖两种组合方式进行砌筑时，具有较大灰缝的边砖和立砖为砖之间提供了更大的粘结强度和更好的压力扩散性能，这样就不会像水平砖块一样容易断裂，故在过梁、支座和檐口中经常采用这种方式。

P18　　　　　**砌体砌筑**

为了利用砌块和砂浆制作出具有优良承载能力的高质量砌体，在铺设砖块时，必须遵循一定的工艺标准——砌筑标准。通常有4种砌筑方式，根据砌块层的铺设顺序和水平错层形式，上述4种方式均很容易掌握。

一些砌筑标准很常规，下面先给出两种常见的砌筑方式。

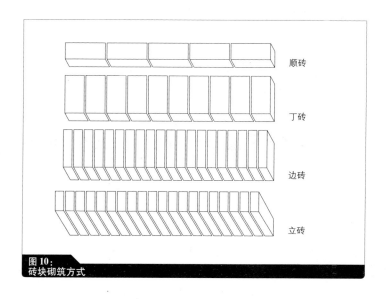

图10：
砖块砌筑方式

规则为：
—所有的砖块均水平铺设；
—砖块高度不大于砖块宽度；
—同层只能采用相同高度的砖块（墙体端部每隔两层除外）；
—尽可能多的采用整体砖块；
—竖向上两层之间水平错层长度至少为1/4砖长。

砌块的水平错缝对于墙体的承载力是至关重要的。水平错缝越大，即砌块的阶梯形齿缝搭接越浅，则砌体抵抗纵向开裂能力越大。

顺砖砌筑

在顺砖砌筑中，砌体中各皮砖均采用顺砖组砌方式，且上下层错缝半砖长。由于此种墙体砌筑方式不允许采用垂直于墙体轴线灰缝的砌筑方式，故仅适用于厚度为半砖厚的墙体，例如内墙、面层和烟囱等。厚度较大的墙体需采用尺寸较大的砖块才能完成。顺砖砌筑中砖块搭接长度较大，因此可提供较大的抗压或抗拉强度。虽然也可以采用1/3或1/4砖长的搭接长度，但此时会导致砌体承载能力的下降。

丁砖砌筑

在丁砖砌筑中，砌体中各皮砖采用丁砖搭接砌筑，且上下层砖搭接长度为1/4砖长。此种砌筑方式适用于整砖厚的墙体。由于搭接（错缝）尺寸较小，此种砌筑方式承载能力低，容易产生斜裂缝。但此种砌筑方式特别适用于半径较小的砌体。

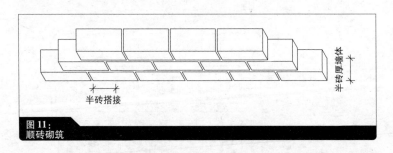

图11：
顺砖砌筑

提示：
此处砖块长度是指标准砖块的长度，但必须考虑砂浆灰缝的尺寸。因此对于专用尺寸为25cm的砖块，1/4砖块长=专用尺寸/4 –灰缝尺寸=25cm/4 –1cm=5.25cm。同样的方法适用于砖块宽度或墙体厚度：两块砖厚的墙体厚度=2×25cm –1cm=49cm（外部尺寸）。

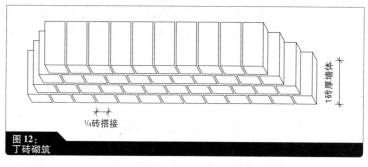

图12:
丁砖砌筑

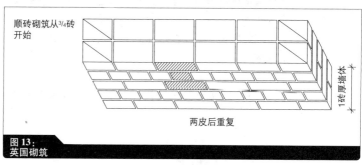

图13:
英国砌筑

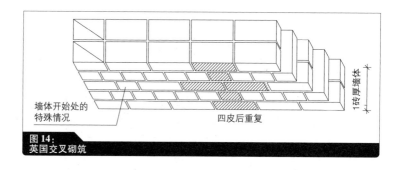

图14:
英国交叉砌筑

我们可以采用上述两种以上的砌筑方式进行组合砌筑。下面列举两种常见的组合砌筑方式。

规则:

—顺丁组合砌筑;

—以3/4砖块开始的顺砖砌筑(对于厚度较大的墙体,相应地采用大于3/4砖的砖块)。

英国砖筑方式

英国砌筑方式为顺砖层与丁砖层交替砌筑方式。两皮砖的搭接长度为 1/4 砖长，这就形成了 1/4 砖长和 3/4 砖长两种尺寸的阶梯形灰缝。

英国交叉砌筑

如英国砌筑方式一样，英国交叉砌筑也由一顺砖层和一丁砖层交替砌筑开始。但顺砖层在竖向以半砖长进行搭接，故以灰缝模式每4层一重复。此种砌筑方式具有灵活多样的灰缝模式，但由于阶梯形灰缝坡度较大，因此极易产生斜裂缝。

也有一些装饰性的砌筑方式，但这些砌筑方式现已不存在或仅在某些局部区域流行，诸如：双层佛兰德砌筑、约克夏砌筑和佛兰德砌筑方式等。

墙体横向上可通过顺丁组合砌筑方式，建造一厚度大于整块砖厚的墙体。

采用的工艺标准如下：

—对于厚度较大的墙体，如有可能，仅采用丁砖砌筑；

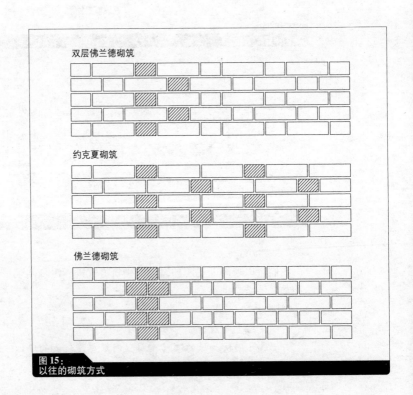

图15：
以往的砌筑方式

—如有可能,砌体竖向灰缝贯穿整个墙厚;
—砌块搭接长度不小于1/4砖长(墙内竖向灰缝);
—墙体纵向和横向,砖块均需进行有效搭接。

P22
墙角

标准施工

对于转角、墙洞、突出部和柱子,有一些特殊的砌筑细节需注意。

工艺标准:
—顺砖皮与丁砖皮交错布置,且贯穿墙角、交叉节点等;
—平行墙体具有相同的砌筑方式;
—每皮砖内仅有一个竖向灰缝起自内墙角;

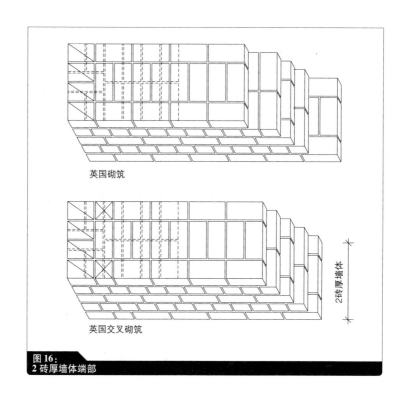

**图16:
2砖厚墙体端部**

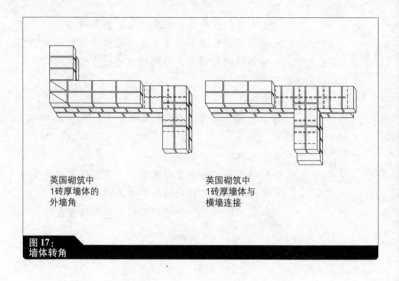

图17：
墙体转角

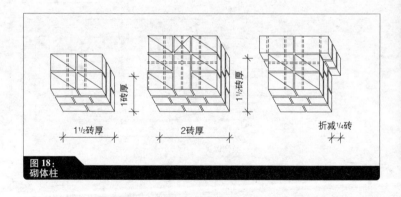

图18：
砌体柱

砌体柱

—窗洞和门洞应如突出的墙体端部一样砌筑，对于丁砖砌筑，在墙体突出方向上用一砖代替；对于顺砖砌筑，连续铺设即可。

进行砌体柱施工时，有两点需注意：

—方形柱各皮的砌筑方式相同，且每次转角90°。

—跟墙体端部类似，矩形柱窄边以3/4砖开始。空隙处填满整块或半块砖。

砌体洞口

墙体及墙体中的窗洞、门洞和出入口应该满足施工工艺标准和传统习惯。

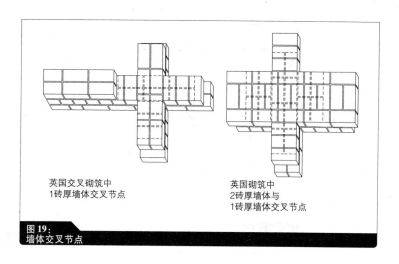

英国交叉砌筑中
1砖厚墙体交叉节点

英国砌筑中
2砖厚墙体与
1砖厚墙体交叉节点

图19：
墙体交叉节点

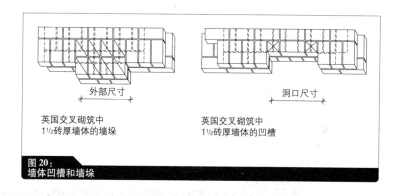

外部尺寸

洞口尺寸

英国交叉砌筑中
1½砖厚墙体的墙垛

英国交叉砌筑中
1½砖厚墙体的凹槽

图20：
墙体凹槽和墙垛

提示：
由于承重墙施工中新型砖块规格和技术的发展，使得承重墙通常采用较自由的砌体砌筑工艺（而非遵循特殊的砌筑工艺标准，但砌块搭接必须满足最小的搭接尺寸），而上述砌筑工艺只有在清水砌体建筑中才会采用（见"砌体结构"之"外墙"）。

提示：
现在，烟囱通常是由特殊用途的砌块建成，因此清水砌体建筑仅进行简单的面层处理即可。这里图示的砌体结构仅仅表明了砌体的砌筑工艺标准和可能性。

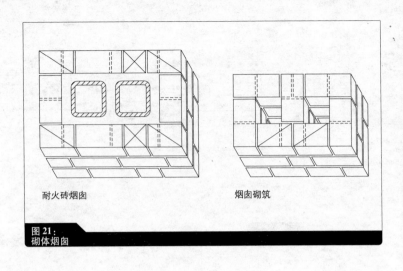

图 21：
砌体烟囱

耐火砖烟囱　　烟囱砌筑

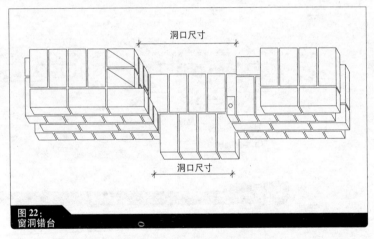

洞口尺寸

洞口尺寸

图 22：
窗洞错台

　　窗洞的侧面错台和门洞应按砌筑工艺标准砌筑，这样既简化了门窗安装，同时又提高了抵抗风雨的能力。

　　门洞和窗洞上的墙体也应按工艺标准施工。由于砌体不能抵抗弯曲内力，洞口上部不能采用没有"支撑"的砌体结构，因此洞口上部应布置过梁，过去通常采用木梁或石梁，而现在通常采用混凝土梁。这些过梁将来自上部砌体的外加荷载扩散传递给洞口两侧的墙体，这样过梁材料的选用就决定了洞口尺寸的大小。

　　适用于砌体的另一种洞口形式是砌体拱结构，此种结构将所有的

砌体拱

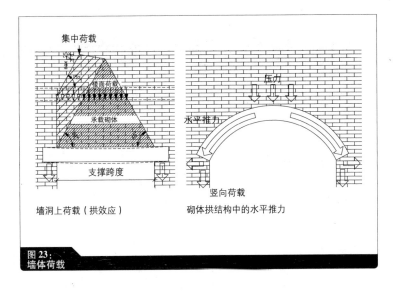

图23：墙体荷载

外加荷载转化成压力，并传递给支座。此结构的难点在于受荷的拱结构对砌体产生的水平推力，该力被墙体或附加的墩柱承担，而且拱结构的矢高越小，推力越大。

圆形拱是半圆形的砌体将外加的荷载传给支座，且支座通常是水平的。因而拱结构的半径为洞口宽度的一半。为了得到此种形状的拱结构，砌块间的灰缝应为楔形，其在拱内弧面的厚度不应小于5mm，在外弧面的最大尺寸为20cm。这就意味着当处理半径较大或跨度较大的拱形洞口时，在拱结构的上部需布置多层砌块。对于半径较小的砌体拱结构需采用楔形砌块。

当拱半径为洞口跨度且拱与下部支座相切时，就形成了尖拱。两种类型的拱由不同数量的砌块构成，因此，需在拱顶布置拱顶石，拱顶石在砌体的层间灰缝处终止，因此拱顶上部填充的砌块层数不会太大。对于尺寸收缩的窗子，可以建在两行砌块之内，垂直变动。

> **提示：**
> 由于砌体的"拱效应"，即将洞口周围的荷载转移出去，因此仅有洞口上部砌体结构自重对洞口上的过梁有影响，即图23所示三角形区域的砌体自重。此外，当集中荷载距离洞口上部三角形区域顶端尺寸不大于25cm时，需考虑集中荷载；当上部楼面荷载在三角形区域内时（见图23），需考虑上部楼面传来的荷载。

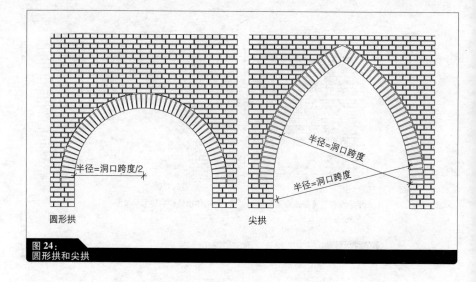

图 24：圆形拱和尖拱

如果周边承重结构可以承担更大的水平力，则可采用浅拱结构。对于扇形拱结构，采用较大半径的圆弧段，此处拱的矢高（拱内侧最高点与最低点的高差）不应大于洞口宽度的 1/12。支座倾斜且指向拱的圆心。

如果洞口几乎水平而仅仅是砖块倾斜，此种拱专业术语称为"平拱"，其矢高最大值为洞口宽度的 1/50。

对于下述两种施工方法，须严格限制洞口宽度，应采用如下规则作为实施标准：

——对于采用 24cm 长砖块的扇形拱其跨度不应大于 1.2m；

——对于采用 25cm 长砖块的平拱其跨度不应大于 0.8m。

砌体拱是一种非常精巧的结构，常见于教堂和著名建筑中，现在已很少采用。目前拱结构作为精巧的结构构件，可采用钢材加工制作而成。

灰缝构造

除了砌体砌筑方式外，砂浆灰缝的构造对砌体的外观也有重要影响。灰缝的颜色和厚度对砌体设计效果起强化作用。

正确处理砂浆灰缝可提高结构的抗力和耐久性能。砂浆灰缝通常有两种形式：

平嵌缝

瓦刀完成的灰缝，当砌块就位时砂浆就从砌块周围挤出，稍后用

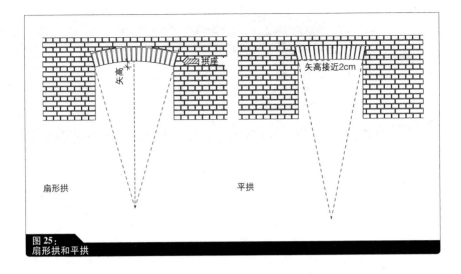

图 25：
扇形拱和平拱

木片或软管清除这些挤出来的砂浆。这种方法的优点在于会对灰缝产生良好的密封性，使砂浆能满足砌体表面外观的需要，同时提高了砌体结构的承载能力。

二次勾缝　　当灰缝的颜色和图案的一致性特别重要时，采用二次勾缝就具有很大的优越性，此时在清理的灰缝表面用木片将新拌砂浆勾抹约20mm厚。对于吸水性砖块，由于砖块将从砂浆中吸收水分，故二次勾缝砂浆在勾缝前需进行加湿处理。由于此方法采用两种砂浆，进行高级装修须慎重使用，以确保砌体的承载能力和密实性。

例如：

对位于芝加哥的罗比住宅弗兰克·劳埃德·赖特采用水平凹缝和竖向平嵌缝进行水平效果强化处理。

例如：

阿尔诺·莱德雷尔对位于斯图加特的办公楼进行了如下色彩设计：选用黑色砌块和竖向黑色灰缝嵌缝，对水平灰缝却采用白色。此处给出了清晰的立面效果图（见图26，左数第二张图片）。

图26:
灰缝

P30

装修工艺标准

砌体需横平竖直地进行砌筑。通常由于砌体的第一皮弥补了地形的不平,因此非常重要;然后紧接着从墙角处开始铺设砌块。对于每块重达25kg的砌块可采用手工铺设,而重量更大的砌块需采用辅助设备。砂浆应均匀布满砌块底面的所有区域。施工时,对于小尺寸砌块,宜采用瓦刀勾缝;对于大尺寸砌块,宜采用砂浆模板处理。砂浆模板可保证砌体整个长度方向上灰缝高度的一致性。采用砂浆抹面或灰缝勾缝,使竖向砌缝需封闭以确保砌体能防风避雨。为了降低工期和节约成本,砌体砌筑时竖向灰缝有时不用砂浆,但此时必须满足保温(采用一层抹灰和覆层)和隔声(优良的隔声砌块)的要求。此时最好采用榫槽体系的砌块。

如果采用高吸水性的砌块,由于砌块将从砂浆中吸收大量的水分而使砂浆变得干燥,因此在铺设砌块前一定要谨慎。同时砌块也将从砂浆中吸取少量的盐分,致使砌块表面形成"盐花"。而且经完全浸泡的砌块不能与砂浆进行良好的粘结。因而砌块和砌体既要避免强光直射,又要防止雨水冲蚀。结霜时,只有采取预防措施才能铺设砌块。此时砂浆的硬化速度随着温度的降低而减缓,当温度达到 $-10℃$ 时,砂浆将停止硬化。当温度降低到5℃时,建筑材料需进行覆盖保温;当温度低于0℃时,砌块和拌合用水需保温。不应采用冻结材料;当部分墙体遭到冻结破坏时,需拆除。

P31

石材建筑

天然石材为乌尔形式的砌体。从尺寸大小不一、不采用砂浆而仅进行简单堆砌的石材,到按照砌筑工艺标准(料石砌体)铺设的尺寸整齐划一的石材,存在着诸多类型的天然石材砌体结构。然而,目前采用天然石材作为砌体实际材料的很少,较多的是作为墙体的幕墙

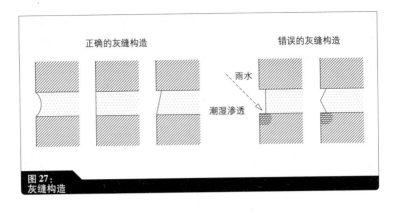

图 27：
灰缝构造

材料，因而这里不做过多的介绍。然而在纪念碑或景观建筑中却经常采用天然石材。

P32

新方法

除了砌体砌筑工艺标准规定的传统建筑方法外，基于新的建筑材料、新的加工方法和新施工方法也得到了发展，以降低施工成本、缩短施工工期。

模制砖砌体

对于模制砖砌体，由于砖块的尺寸偏差减小了，致使灰缝的厚度减低到 1~3mm（薄垫层）。砂浆采用滚筒施加，或将砖块浸于水中。由于灰缝厚度减小和砌块一致性的增大，既节省了材料和时间，又获得优良的承载能力，而且降低了热桥效应（详见"砌体结构"之"结构特性"）。

由于仅允许砌块有较小的尺寸偏差，故第一层砌块的铺设需慎重。基于此目的，砌块的错缝尺寸要减小。在不同的高度可采用具有优良绝缘性能的较小错缝的砌块。

无砂浆砌体

无砂浆砌体不使用砂浆。然而由于承载力的原因，这种砌体结构只适用于低层建筑和高度较低的建筑。墙上部楼面荷载需均匀，这样楼面荷载产生的压力就可抵消砂浆粘结力缺乏造成的不利影响。

提示：
　　热桥作为保温隔热的薄弱点，导致了建筑物热量的流失。如果建筑的吸热面积小于散热面积，且采用了不同的材料，结构上采用传热构件和隔热构件，这样我们就可以确定热桥效应（例如建筑的转角处）。

砌体施工工具　　　　为了节省切割石材所需的时间，工厂里砌体施工工具为我们提供了生产合适尺寸墙体的可能，并将加工好的整片墙体运送到现场。此方法价格合理，特别适用于山墙或带洞口的墙体。

预制砌体方法　　　　此种方法是在工厂阶段提前预制砌体，制造商将带有洞口的整片墙体运送到现场。运输过程中需对砌块加固以使结构稳定，安装时需要动用起重机或移动式起重机。尽管成本相对较高，但由于是工厂加工构件，质量稳定可靠（尽管在现场安装工人要负责墙体连接节点安装），弥补了成本的不足。

P35　　## 砌体结构

下面列举的结构指的是处于施工状态的墙体。前面阐述的施工工艺标准原则上适用于所有的砌体结构，并且仅是处理砌块与砂浆的砌合问题。实际上有许多种依赖于建筑材料和建筑构件的施工方法，这些方法取决于建筑物所处的位置和墙体结构的作用。

砌体墙体能够承受来自上部楼面的竖向荷载、自重荷载和水平荷载，诸如风荷载、土压力和冲击荷载，或者来自突出构件或悬挂构件的悬臂荷载。

基于上述原因，墙体必须与相邻的建筑构件相连接，即荷载需经其他的承荷构件传递和转移或直接传递给地基基础。采用联结墙或在墙体上施加均匀的竖向荷载稳定墙体，避免墙体失稳。确定墙体的尺度时，更多的还要考虑满足其防火的需要。仅承受本层自重荷载和水平力的墙体可作为非承重墙建造。

P35　　　　**结构特性**

砌体的承载能力由砌块和砂浆的粘结状况确定。砌块和砂浆之间的粘结和摩擦影响了水平力的传递，并将竖向荷载均匀分布在整个区域上，同时灰缝抵消了砌块的尺寸偏差。灰缝承受压力的能力远大于承受拉力或弯曲拉力的能力，即采用砂浆垫层砌筑后的砌块能传递压力荷载，否则砌块可能会断裂。砌块和砂浆灰缝受到来自上部的竖向荷载作用而压合在一起。砌块传递压力，而砂浆由于更易变形，不断被压缩。这些不同的特性导致在砌块和砂浆的接触面上产生应力，砂浆上为侧向压应力，砌块为侧向拉应力。与此同时，砌块上的侧向拉应力降低了砌块的抗压强度。随着荷载的增大，砌块上就会出现竖向

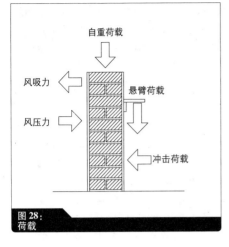

图28:
荷载

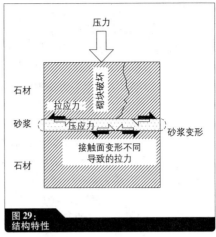

图29:
结构特性

裂缝,并且砂浆将被压溃。如果砂浆不平整,就会增大砌块上的侧向拉力,加大砂浆压溃的风险。砂浆砌缝的厚度越大、砂浆自重越轻,砌缝的变形也越大,因此上述压溃的危险越大。比重大的砌块具有良好的传力性能。

加气砌块和孔洞削弱了砌块的横截面,降低了砌块的承载能力。砌块和砂浆之间的粘结力也允许力沿水平方向传递。

砌块的抗压强度是另一个重要因素。砌块和砂浆须相互匹配,以避免灰缝压溃。对于砌块和砂浆需给出抗压强度等级作为两者的特征值。

P36

外墙

其他承重体系(框架结构、结构板等)中的填充墙或独立墙体除外,砌体外墙一般均为承重墙。外墙将建筑内外空间隔开,这样就可以避免外界严寒、雨、雪和噪声的侵袭。同时设计方案决定了砌体的建筑效果。

单叶砌体

仅连接一片墙体的外墙称为单叶砌体。此种结构工艺上易于安装,但却具有外墙的全部功能。

提示:
相对密度为质量与体积的比值。由于其大小随着吸收水分的增多而增大,因而通常根据干燥砌块确定,即干密度,单位 kg/m^3。

单叶清水砌体

单叶清水砌体，一种从两侧或至少从外侧不需要装修即可达到建筑效果的墙体结构，对于保温隔热和防止气候侵袭两种情况性能差异很大。为了达到目前的保温隔热标准，需采用具有良好隔热性能的加气砌块。由于静止的气体具有较低的导热能力和较小的密度，故孔洞内充满大量气体的砌块相对密度小，却具有良好的保温隔热性能，然而避免气候侵袭的能力却较弱。

当砌块的毛细孔里迅速充满潮气时，砌块就不能抵抗霜冻的侵袭，故不应对其未加任何保护措施而直接使用；相反，能较好抵抗气候侵蚀的砌块通常具有较高的相对密度，但保温隔热性能却较差，故需更大的墙体厚度，因而是不经济的。此种墙体目前已不再使用。

饰面砌体

另一方面，对于饰面砌体，墙体由不同厚度的单元构成。外侧砌块具有良好的避免气候和霜冻侵袭的功能，而内部单元负责保温隔热。此处，包括饰面在内的整个横截面承担荷载作用。抗压强度最小的砌块提供了砌体承载力计算依据。两套石材之间的 2cm 厚的密封砂浆灰缝、错缝给内部单元提供了保护。这种饰面砌体是一种精巧的结构，内外单元材料要相互匹配。以避免产生不同的沉降率和变形。对此应进行准确的规划设计，以避免单元规格不协调。这里建议仅对强调视觉效果和有正式要求或采用特殊砌块或当清水砌体中不设置伸缩缝时，才采用此种结构（见"砌体结构"之"外墙"）。

由于上述所有特点相互依赖，对于单叶砌体需采取附加措施以使其免受气候的侵袭。

采用外抹面的单叶砌体

例如，我们可采用外抹面以提高保温隔热性能。尽管在单叶砌体中，采用外抹面会丧失砌体砌筑的视觉效果，但可采用大幅面的单元，使用较薄的砂浆层自由砌筑。大幅面单元具有优良的隔热性能，而且使用起来比较经济。由于墙体整个横截面对保温隔热均有影响，因此应采取措施以避免产生热桥。特别是对于过梁和顶棚支撑等，需要采用特殊的施工工艺。

提示：

相对导热系数指的是在确定条件下，一个结构构件传递的热量。其值越小，构件的保温隔热性能越好。

提示：

热传递阻抗（R）是指基于结构构件厚度的隔热能力，其值为结构构件层厚度与相对导热系数的比值，也可计算结构构件在极端条件下的热转变及多层结构构件的热传递阻抗值。

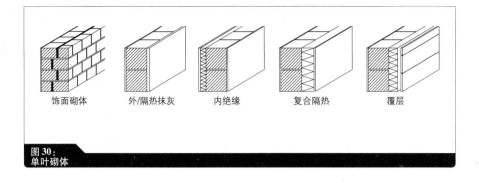

图30：
单叶砌体

饰面砌体　　外/隔热抹灰　　内绝缘　　复合隔热　　覆层

顶棚支座

　　顶棚由墙上支座和墙体产生的粘结力和摩擦力与周围墙体相连，通常需采用 10～12cm 厚的支撑边。

　　由于钢筋混凝土的热传递阻抗小于砌体的热传递阻抗，顶棚支座降低了保温隔热性能，从而导致在顶棚和墙面上形成低温区，使来自内部温暖空气中的湿气在墙体内表面浓缩凝结。

　　基于上述原因，顶棚的外边需提供附加的保温隔热措施。此处需要注意的是，由于不同材料在接触面处的膨胀和变形不同，单叶抹面墙的外侧容易产生裂缝，从而导致雨水渗入建筑内。为了避免此种状况发生，可采取构造措施，连接过渡区域以确保抹面，也建议采用 L 形骨架过渡区。这些均由与墙体相同的材料构成，而且有些已经布置了隔热带。如同钢筋混凝土框架构件一样，从而避免了材料变化。

提示：

　　如果没有足够的面积支撑顶棚，砌体中需埋设拉接钢筋。这样在砌体中就会产生水平拉力，墙体需要一个相应的外加荷载以抵消拉力。这就意味着拉接钢筋不能固定在护墙内。山墙也可采用拉接钢筋附着在屋盖结构上。

例如：

　　冷凝物：温暖的空气较寒冷的空气包含更多的水蒸气。当冷、暖空气相遇时，就会形成薄雾或水蒸气。当暖空气遇到寒冷的物体时，就会产生大量的水分，形成冷凝物。对于房间内的热空气，当外墙的保温隔热性能较差，或根本不能保温隔热时，水分就会在冰冷的内侧面或结构构件上冷却沉淀。这就导致结构发生冻融破坏或霉变。

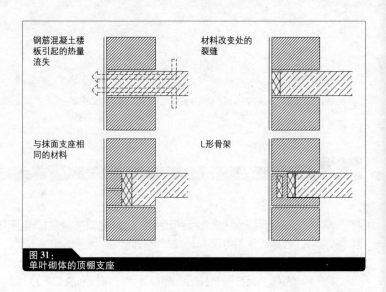

图 31：
单叶砌体的顶棚支座

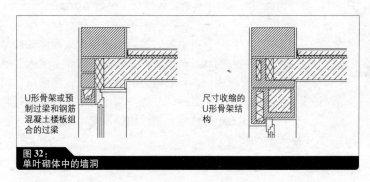

图 32：
单叶砌体中的墙洞

墙洞　　墙洞中也会发生上述类似情况。由于砌体不能承受拉力或弯曲荷载，因而不采用支撑结构而直接建造墙洞是不可能的，需要外加过梁以抵抗荷载作用，并将荷载横向传递给相邻的墙体。当采用钢梁不能

> 提示：
> L 形或 U 形骨架结构可由砌块制造商预制。正如名称所述，L 形骨架结构为 L 形，以支撑顶棚；U 形骨架结构为 U 形，应用于墙洞上部以形成圈梁。在现场，两者的孔洞中充满了混凝土（见图 31）。

满足防火标准时，像楼面板一样，通常采用钢筋混凝土梁，而且要具有附加的保温隔热措施，或建造成U形骨架结构。这些特殊的构件可以现场制作，即与混凝土楼板同时浇筑，或在工厂预制，运输到现场。

连系梁/圈梁

我们也可以采用U形骨架结构以形成外围的连系梁和圈梁。其他因素，诸如风荷载，使建筑物产生了拉力。这些荷载被刚性楼板传递，且不能只由墙体承担。外围的连梁由钢筋混凝土梁或楼板下的U形骨架结构或钢筋混凝土板带组成。他们将所有的荷载传递给外墙和横墙。当楼板为非刚性楼板或有滑动支座时（例如当采用平屋顶时），外围的连梁应沿着整个建筑连续布置成环形（圈梁）。

内部隔热的单叶砌体

为了提高墙体结构的保温隔热性能，可在墙体内侧采用绝缘层或保温隔热层。根据建筑科学理论，这种结构存在问题，诸如：由于冷凝水会在低温砌体内侧形成，且湿气会渗入结构内，从而导致霉变。基于上述原因，对于二次改造且不允许改变建筑立面效果的结构，可以采用内部隔热的单叶砌体。

层状隔热体系（LHIS）

为了避免上述问题，在层状隔热体系中，绝缘层并非布置在墙体的内侧，而是粘贴在砌体上并采用拉条固定。为了使绝缘层免受天气的侵蚀，采用一层特殊的防水、防潮层作为绝缘材料。当抹灰层需要实心底层时，且外界因素没有形成空洞或压力点时，绝缘层需能抵抗压力且能提供充足的抵抗能力。当考虑经济原因时，层状隔热体系是一种普通的体系，特别是对改造旧有建筑时更是如此。

带覆层的单叶元砌体

另外一种保护承重墙体的方法是在建筑上悬挂外壳，该外壳由金属、木材或纤维水泥构成，直接粘贴在砌体表面，或留有空隙以布置附加的绝缘层。对于悬挂固定点需注意，固定点会冷却砌体，并且应进行充分的通风以阻止来自覆层后面的水分形成的湿气渗入。

地下室墙体

所有结构中的地下室墙体均为单叶砌体。目前采用防水钢筋混凝土的方法已经越来越普遍，但也可以采用其他的墙体结构。地下室墙体需进行加固以抵抗土压力和荷载传递，同时土压力影响了墙体表面。当确定地下室墙体尺度时，需要考虑墙体高度、土压力和来自地表的外加荷载。对于有较高隔热要求的地下室施工区域，地下室外墙外侧采用的隔热层需能抵抗土压力作用。隔热层由泡沫玻璃纤维布、聚苯乙烯粒状泡沫或挤压的聚苯乙烯泡沫板组成。地下室墙体必须封闭以防止土壤中湿气侵入。当水的压力势较大时，就要采用混凝土抹面；当荷载较小且没有水压力时，需采用水平和竖向封条。在混凝土

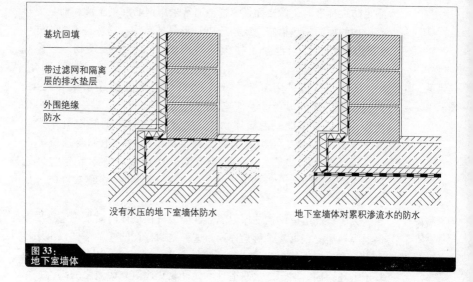

基坑回填

带过滤网和隔离层的排水垫层

外围绝缘防水

没有水压的地下室墙体防水　　　地下室墙体对累积渗流水的防水

图33：
地下室墙体

板的外表面，应采用水平薄膜，该薄膜为密封层，同时必须与第一皮砖下砌体外侧采用沥青覆层的竖向封闭层相衔接。混凝土涂层提供了附加的保护：此处水平薄膜附着在楼板下的底层砌块上，并有一层保护涂层。基坑回填时，应采取措施保护竖向薄膜和隔热层以防止遭到土壤破坏，诸如采用土工织物薄膜保护层和过滤网保护层。

勒脚区域

相对上部砌体来说，勒脚区由于相邻土壤和水飞溅效应的作用将承受更大的荷载，因而采用地表上30cm高的竖向薄膜进行封闭以防止潮气侵入，并以整个墙体宽度的水平防潮层结束。该水平防潮层可避免潮气上升进入上部砌体。它可通过采用防天气侵蚀的砌块、覆层或特殊的防水勒脚抹面加以保护。抹面之间的过渡可通过构造或采用不同的光滑度实现。例如膨胀金属中的抹灰底层，可避免裂缝产生。

双叶砌体

在双叶砌体中，保护内侧面免受气候侵蚀的第二片墙体（外墙或饰面叶）建于内墙的前面（内叶），同时内墙具有基本的承重功能。第二片墙体与内墙间的空隙被保留下来，其间或不填充，或部分或全部填充绝缘材料。

带孔洞的双叶砌体

此处布置孔洞是为了阻止水分直接渗入内叶砌体，进入建筑内部，造成诸如霉变等破坏。如果湿气渗入外叶砌体，就可通过孔洞清除。通风孔应布置在勒脚区、墙体上部和墙洞处。这些通常是带水平防潮层的竖向敞开灰缝。通过在敞开的沉降缝下布置密封条等作为Z

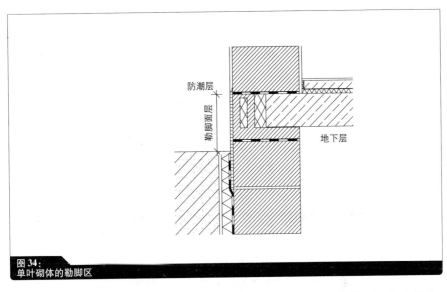

图34:
单叶砌体的勒脚区

形栅栏穿越水平灰缝的整个区域,通过1~2cm的斜面与内叶相连,并向上延伸15cm,实现上述情况。

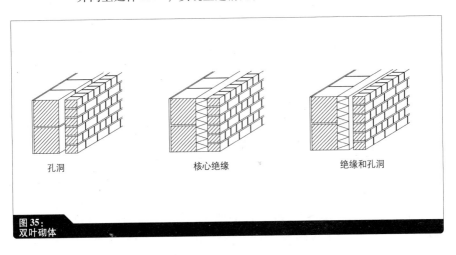

图35:
双叶砌体

> 重要提示:
> 对于所有的密封薄膜,凡是卫生导管或服务节点穿越墙体或顶棚的节点,均需格外注意,这些节点需仔细密封。

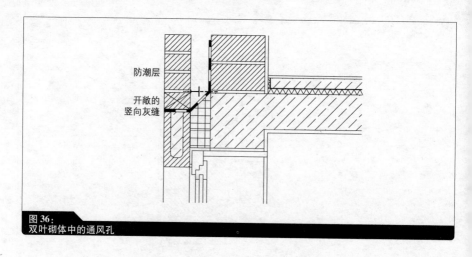

图36：
双叶砌体中的通风孔

为了确保充分的通风效果，通风孔不应小于60mm；当灰缝砂浆清除干净或采用绝缘层时，则不应小于40mm。尽管竖向通风孔，包括后部通风，几乎不传热，但为了更好地保温隔热，在通风孔中安装绝缘层通常是必要的。如果两叶之间的通风孔填满了绝缘材料，则称为带绝缘核心的双叶砌体。

完全填充
墙体

通过整个墙体厚度，而非砖块厚度增大了墙体抵抗热传递的能力。绝缘层呈带状或条状固定在内叶或松散的颗粒或混合物上。松散的颗粒或混合物被填充在空隙内，填充时需注意，一定要分布均匀。此种结构的缺点在于水分能进入前叶的后面，而且很难清除。由于潮湿建筑材料的导热性能大于干燥建筑材料的导热性能，故水分的渗入降低了结构的隔热性能，因而绝缘材料须能长期防水、灰缝、连接节点须能阻止水分渗入。通常采用裹紧的软矿物纤维布、塑料泡沫或舌榫结构等措施阻止水分渗入。由固定绝缘层或外叶造成的破坏应进行密封。如果两叶之间的空隙中填充了绝缘材料，此时必须注意：通过安装防腐栅格，避免材料落入排水孔中。

带有绝缘层和通风孔的双叶砌体具有上述两种结构的优点。

部分填充
墙体

防水隔热层附着在内叶砌体上，并通过不小于4cm的通风孔与外叶砌体隔开。这种结构比其他结构更精巧。由于承重、绝缘、防潮和气候保护被严格区分开，因此此种情况具有优良的性能，然而整个墙体厚度将变大。

内叶

在所有的结构中，内叶砌体主要保持结构稳定性和传递荷载。内叶砌体可采用相对密度较高的承重砖块建造。这些砖块抵抗热传

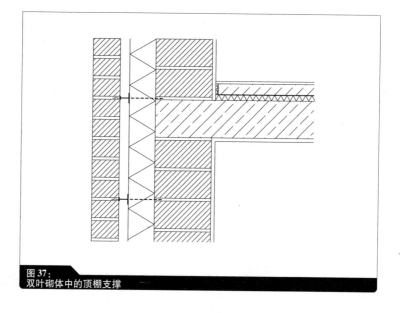

图37：
双叶砌体中的顶棚支撑

递的性能较差，但却具有较高水平的隔声效果。实际上，对于内叶砌体可采用建设部门批准的所有标准砌块和砂浆（见"建筑材料"）。由于内侧面通常有一层抹面覆于砌块上，采用大型砌块尽管违反了砌体砌筑标准，但由于大型砌块采用自由砌筑方式砌筑并采用较薄的砂浆层，具有很大的强度。内叶砌体前的隔热需连续布置。顶棚须由内叶砌体支撑，通过在内叶砌体上合理地布置绝缘带可附带提高隔热性能。

外叶　　外叶砌体保护砌体免受外界因素和气候的侵蚀。基于此原因，只有满足下述条件并对霜冻、潮湿和外界条件影响不敏感的材料才能采用。这些材料包括砌块、硅酸钙和混凝土砌块（见"建筑材料"）。砂浆制作商可提供特殊的防霜冻砂浆，此种砂浆几乎不吸收水分，风化程度低，即经盐渍沉淀后不褪色。

外叶砌体决定建筑物的外观，可采用上述砌筑工艺建造。但外叶砌体只能承受自重，并采用锚固钢筋固定在内叶砌体上，以确保外叶砌体能抵抗风压力或吸力作用，以避免倾覆、倒塌或翘曲。所需锚固钢筋数量和直径取决于内外叶墙体间距和墙体高度。对洞口敞开边缘、建筑拐角处、伸缩缝以及结构中的圆形构件需特别注意：应采取合适措施以防止潮气从外叶墙体进入内叶墙体，诸如安装塑料圆板以使水分从孔洞中排出。

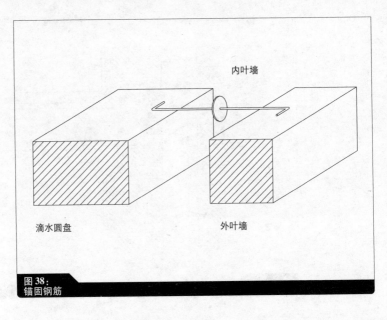

图38：
锚固钢筋

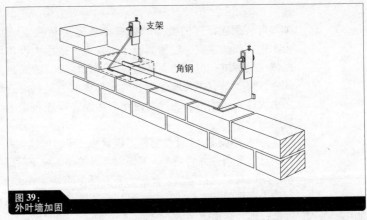

图39：
外叶墙加固

支撑加固　　除了对墙体锚固固定外，外叶墙需支撑、附着在内叶墙上，这样当外叶墙体高度较大时，其自重就可均匀地传递给承重墙叶和基础。对此，可采用防腐锚固支架、角支架或隔热顶棚凸出部加以解决。

应采取措施以避免底部发生滑移，锚固支架布置位置应尽可能低。低处的密封带应延伸到外叶墙的前边。

外叶墙的最小厚度为9cm，低于9cm的称为墙体覆层（见"外

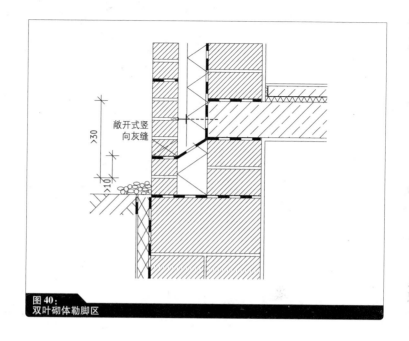

图40：
双叶砌体勒脚区

墙")。考虑空间和成本原因，外叶墙通常采用半砖厚，因而砌筑方式是不规则的。大多数普通的砌筑方式（除了顺砌外）均不能这样施工。

洞口　　基于审美原因，砌筑总是遍布墙体整个区域，因而洞口、窗户、门及任何突出构件均需进行特殊锚固就位。立砖砌筑中经常采用过梁，与拱结构不同，它不是普通结构，不能承受荷载，故砌块须由托架支撑。采用这种方式成本较低，从外侧看是可见的，也可采用不可见的灰缝加固以使砌块就位。砌块制造商提供已加固并填充混凝土的U形骨架结构。这些结构将荷载传给两侧相邻的墙体。所有的金属构件均应是防腐的，最好采用不锈钢。因为镀锌构件容易在运输或安装过程中受到破坏，且这些缺陷很难发现，安装后难于处理。

变形缝　　由于受温度和天气影响程度不同，外叶墙体的变形不同于内叶墙体的变形。对于外叶墙体应布置水平变形缝和竖向变形缝以承受上述变形。与材料相关的伸缩缝间距（见表1），基于与界限点相关的因素，将墙体在转角处分开。西墙变形最大，北墙变形最小。考虑建筑上的需要，这些变形缝应向内错开变形缝间距一半的长度。通过在窗台两

103

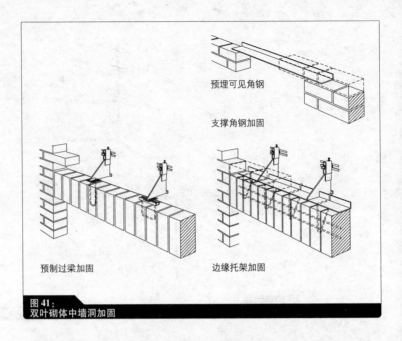

图 41：
双叶砌体中墙洞加固

侧设置伸缩缝，避免由窗台荷载和窗台周围砌体荷载产生的窗台周围裂缝，也可对窗楣进行结构加固以代替这些伸缩缝。在这些加固托梁下通常布置水平变形缝。

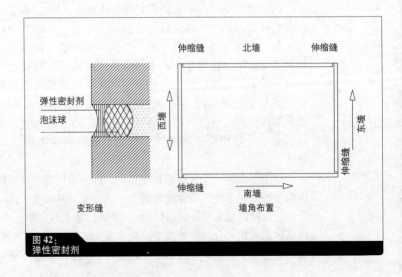

图 42：
弹性密封剂

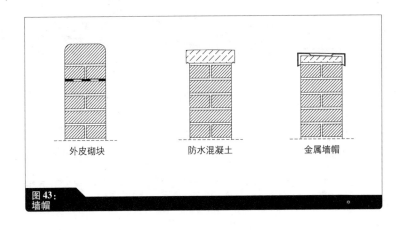

图43：
墙帽

表1：
变形缝间距

砌体类型	伸缩缝间距（单位：m）
硅酸钙砌块、加气混凝土砌块、混凝土砌块	6~8
轻质混凝土砌块	4~6
普通砖	10~20

引自：P. Schubert: Zweischalige Aussenwande – Dehnwngsfugen in der Aussenschale (Verblendschale), in: Mauerwerk 6/2003, Ernst & Sohn, Berlin, P. 203.

非承重外墙——独立式墙体

由于独立式墙体仅由基础支撑而没有外加荷载使其稳定，故墙高限制极其严格。独立式墙体需增加厚度或采用交叉墙体、交叉柱加固以使其稳定。由于独立式墙体位于室外，容易受霜冻的侵蚀，因而独立式墙体必须采用抗冻融的材料和基础，并采取相应措施以免受潮湿的侵蚀。在大地水准面上需布置水平防潮层，墙体顶部需采用砌块、金属板、混凝土墙帽和防潮层加以保护。

相同砌块不同墙体结构的比较如图44所示。给出的热传递系数越小，则其隔热性能越好。图中给出的相互关系比精确的数值更有意义。

内墙

内墙不能直接与室外相连，通常需要借助外墙、顶棚和楼板的保

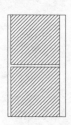 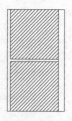 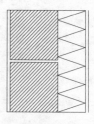

外抹面墙体

轻质矿物抹面2cm 0.31W/m·K
横向砌块30cm 0.14W/m·K
内抹面1.5cm 0.7W/m·K

U=0.417 W/m²·K

隔热抹面墙体

隔热抹面3cm 0.07W/m·K
横向砌块30cm 0.14W/m·K
内抹面1.5cm 0.7W/m·K

U=0.362W/m²·K

复合隔热体系墙体

复合隔热体系6cm 0.035W/m·K
横向砌块30cm 0.14W/m·K
内抹面1.5cm 0.7W/m·K

U=0.234 W/m²·K

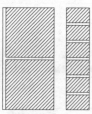 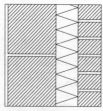 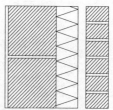

空隙墙体

饰面层11.5cm 0.68W/m·K
孔隙、锚固钢筋4cm
横向砌块30cm 0.14W/m·K
内抹面1.5cm 0.7W/m·K

U=0.412 W/m²·K

绝缘核心墙体

饰面层11.5cm 0.68W/m·K
绝缘核心6cm 0.35W/m·K
横向砌块30cm 0.14W/m·K
内抹面1.5cm 0.7W/m·K

U=0.236W/m²·K

绝缘核心和空隙墙体

饰面层11.5cm 0.68W/m·K
孔隙（锚固钢筋）4cm
绝缘核心6cm 0.35W/m·K
横向砌块30cm 0.14W/m·K
内抹面1.5cm 0.7W/m·K

U=0.242 W/m²·K

图44：
墙体结构类型比较

护，使其免遭严寒、雨雪等的侵蚀。内墙的功能为分隔内部空间、功能区或观光区，诸如住宅、卧室和生活区之间，办公区和生产区之间等，这些隔断需要很好的隔声效果，或者具有良好的防火功能。

一些内墙也要承担建筑物的部分荷载，增大建筑物刚度或增大墙体截面。内墙或与相邻结构构件直接相连承受荷载，或不承受荷载而将自重荷载和水平荷载传给其他结构构件。这些都反映在对连接节点尺寸和细部构造的确定上。内墙的相对密度不但影响其抗压强度，更影响隔声效果。具有较大质量和相对密度的内墙具有优良的抗压强度和良好的隔声效果。

加劲承重内墙　　承重内墙增大了建筑物的刚度，并为顶棚提供支撑。为了增大墙体刚度，墙体连接件应能抗拉或抗压。为了维持墙体稳定，应尽可能选用变形性能类似的建筑材料。将两片墙体建造成相同的高度或预留孔洞（套接）或在墙体中伸出砌块形成连接节点，接着安装加劲墙体。这样做的优点是节点可在后续中发挥作用，只是有时需要额外的空间，例如安装脚手架等。然而，此种方法需要在砌缝中埋设钢筋以承受拉力。

一个有效的方法是采用对接砌筑，这也需要抗拉钢筋或锚固钢筋；此种节点的做法将以后介绍。这种连接只有在墙角的内拐角处才能采用，其优点为：当与外墙相连时，外墙的隔热性能不会因内墙互锁砌块的存在而降低，此时内外墙可能由不同的材料构成。

公用墙体　　基于隔声目的，分隔住宅的墙体通常采用双叶砌体结构。空隙宽度由分隔墙叶的质量决定，此处建议采用5cm。空隙内应填满扎紧的矿物纤维，并在表面覆上矿物纤维布。通过滞后两层错层，进一步提高隔声效果。不允许采用刚性泡沫材料，同时还要注意：不允许砂浆落入灰缝。当施工顶棚时，在墙顶或顶棚处，绝缘层通常应连续布置。

非承重内墙　　对于加劲内墙或传递荷载的内墙，通常不会采用非承重墙体。非承重内墙不能承受风荷载，仅能承受自重荷载和较轻的支架荷载（例如：隔板架、图画架等），并能将冲击荷载传递给相邻的结构构件。非承重内墙长度根据其高度、与相邻结构构件的附着方式（从两边到四边附着）及顶棚变形产生的外加荷载，经计算后确定。

非承重内墙与相邻结构构件的连接节点既可以是刚性的，也可以是滑动的。当没有来自其他结构构件荷载产生的直接应力作用时，可采用刚性节点。刚性节点具有良好的隔声效果和防火性能，同时由于刚性节点没有抹灰，且仅是嵌入钢构件，因而从外表看不明显。

提示：
热传递系数为所有热传递系数和的相反值。

重要提示：
对于竖缝未勾缝的砌体需进行专门计算。

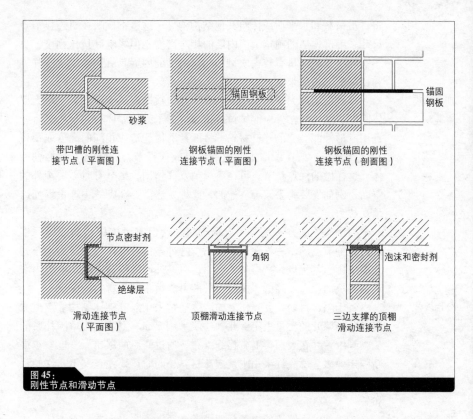

图45：
刚性节点和滑动节点

滑动连接节点采用型钢或滑动锚固节点制作，并能承受一定的变形。这些连接节点非常巧妙，并且是可见的，有时需要掩饰。

P54
凹槽和洞口
确定墙体尺寸时应注意：一定要根据洞口和孔口尺寸综合考虑墙体横截面尺寸，例如：对于机电设备或卫生设施安装，一定不要超过规定的限值。许多制造商供应带有安装洞口的特殊砌体构件。

P56
建筑材料

前面的章节概括地介绍了砌体结构构件，并根据尺寸、几何形状和应用位置加以区分。现在的问题是何种材料适合何种结构及其承担的功能是什么。下边列举了不同类型的砌块和砂浆以示区别。

砌块类型

除了多种类型的天然石材外，还有大量的人造砌块。此处对天然石材不加以详细介绍，因为在建筑结构中已很少采用天然石材作为纯粹的砌体材料。为了实现砌体的功能，诸如：承重、分隔、饰面、绝缘和防护等，人造砌块以多种方式制造，具有多种不同的性能。基于砌块相对密度与应用条件的关系，进行了如下概括：

较高的相对干密度 = 良好的抗压强度
较高的相对干密度 = 良好的隔声效果
较低的相对干密度 = 良好的隔热性能

砌体标准：欧洲一体化产品标准

欧洲一体化产品序列 EN771（砌体规范）在欧盟国家通用。该序列规范包括：

EN771-1　黏土砖砌块
EN771-2　硅酸钙砌块
EN771-3　骨料混凝土砌块
EN771-4　蒸压加气混凝土砌块
EN771-5　料石砌块（密实、轻质砌块）
EN771-6　毛石砌块

上述标准对砌块原材料、加工制作、应用条件、使用说明和试验测试等提供了详细的说明，却没有规定精确的尺寸大小、名义尺寸等。为了满足贸易需要，这类范畴的结构产品需携带 CE 标签以示符合 EN771 系列标准。产品审批、应用许可仍为一个国家的责任。

标准砌体

黏土砖

黏土砖是世界上最古老的人造建筑材料之一。4000 年前古印度的海若帕城（Haruppa）已经生产黏土砖，其尺寸和形状与今天的标准黏土砖大致相同。首先，泥砖在阳光下晒干，接着烧结的黏土砖转变成一种高技术的产品。由于采用可燃性骨料制作，因而黏土砖内部多孔并具有优良的隔热性能，并能保护烧制温度达到烧结点的构件。砌体的规格、表面形式和材料得到进一步的发展。目前出现了多种不同形式和性能的砌块。黏土砖既代表具有悠久历史的传统建筑技术，又是一种经济的、与时俱进的建筑材料。黏土砖由黏土和肥土混合，经模压、挤压，然后切割成合适的尺寸烧制而成。

欧洲一体化标准 EN-771 对 LD 和 HD 型砖加以区别,并将它们分成 I、II 两种类型,这样就可确定误差限值以维持其抗压强度和质量。对于类型 I,公布的抗压强度的偏差概率应不大于 5%。类型 II 中的所有砌体产品,德国标准已不允许使用。

LD 型砖主要应用于双叶砌体结构的承重内叶中,或应用于抹底灰的单叶砌体。LD 型砖具有较小的干密度($<1000kg/m^3$),并且具有良好的隔热性能。这些可通过在黏土砖中添加聚苯乙烯颗粒或锯屑,烧制后留下毛细气孔来实现。只有在防止渗水的砌体中方可采用 LD 型砖。

总干密度大于 $1000kg/m^3$ 的 HD 型砖适用于防护砌体和无防护砌体。这既包括外叶防护砌体,又包括有良好隔声性能的内墙砌体。

对于上述类型的砌体,区分如下:

实心砖　　　　具有竖向气孔的 HD 型砖,孔隙所占的最大比例
　　　　　　　为底面面积的 15% 或体积的 20%。

图46:
黏土砖与黏土砖砌体

提示:
　　在德国,所谓的用户标准即为将 CE 的分类标准转变为适合本国的标准。这些用户标准对颁布的标准(DIN V 20000-401 到 DIN V 20000-404)给出了精确的数值或容许环境。由于上述欧洲标准没有给出砌体材料的应用条件,这一点在德国的使用标准中也没有涵盖,因而为了完整,补充了一些附加标准。

DINV　105-100　　特殊性能的黏土砖砌块
DINV　106-100　　特殊性能的硅酸钙砌块
DINV　4165-100　　蒸压加气混凝土砌块——特殊性能的高精度单元和构件
DINV　18151-100　轻质空心混凝土砌块——特殊性能的空心砌块
DINV　18152-100　轻质实心混凝土砌块和混凝土砌块——特殊性能的实心砖和实心砌块
DINV　18153-100　混凝土砌块——具有特殊性能的砌块

这些标准对产品性能、特点和区别进行了规定,并以表格的形式记录了具体数值和规格,例如:抗压强度等级、相对密度等级和孔隙率等。

表2：砌块			
材料：	黏土、肥土、含黏土的土壤		
骨料：	锯屑、聚苯乙烯颗粒（可选的）		
制作：	模压与烧制		
尺寸	单位：mm（例如：240×300×238），DF 的倍数（如：10DF）		
砌块类型		强度等级*	密度等级*
LD 型砖	竖向成孔砖	6~12	0.7~0.9
	绝热砖		
HD 型砖	实心砖	8~28（36**）	1.6~2.2
	竖向成孔砖	8~20（36**）	1.2~1.6
	实心饰面砖	8~28（36**）	1.8~2.2
	竖向成孔饰面砖	8~28（36**）	1.2~1.6
	实心工程砖	28	1.8~2.2
	竖向成孔工程砖	28	1.8~2.2
	实心工程砖	60	1.8~2.2
	竖向成孔高强工程砖	60	1.8~2.2
	板砖		

注：* 普通等级
** 高强度砖块数值或工程砖数值（没有专门的缩写词）。

竖向成孔砖	竖向开孔占底面面积 15%~50% 的 LD 型砖或 HD 型砖。此处对开孔类型 A、B、C 和 W 进行了区分。
隔热砖	具有较高隔热性能和特殊成孔类型的 LD 型砖。
实心竖向成孔饰面砖	满足上述成孔规格且抗霜冻侵蚀的一类砖。
实心竖向成孔工程砖	表面涂釉的 HD 型砖。该砖只能吸收少量水分，具有至少 28 级的抗压强度，具有抗冻融性能和较高的相对密度。此处根据上述标准，对实心砌体和竖向开孔具有 A、B 和 C 型孔的砌体进行了区分。高强度工程砖应具有不小于 36 级的抗压强度。
高强度工程砖	具有不低于 60 级的抗压强度和 1.4 的相对密度，具有特殊的抗力性能和耐久性能。
板砖	砌筑钢筋砌体时，利用沟槽来容纳砂浆和混凝土。

图 47：
硅酸钙砌块砌体

表 3：
硅酸钙砌块

材料：	石灰、砂子（石英砂）、水	
骨料：	染料和添加剂	
制作：	搅拌、模压与高压下硬化	
尺寸	单位：mm（例如：240×300×23），DF 的倍数（如：10DF）	
砌块类型	强度等级*	密度等级*
硅酸钙实心砖	12～28	1.6～2.0
硅酸钙成孔/空心砖	12～20	1.2～1.6
硅酸钙饰面砖，实心砖	12～28	1.6～2.0
硅酸钙饰面砖，实心砖	20～28	1.6～2.0
硅酸钙饰面砖/成孔砖	12～20	1.4～1.6
硅酸钙饰面砖/成孔砖	20	1.4～1.6
预制硅酸钙砖		
预制硅酸钙构件		

注：* 普通等级

附加规定涉及搬运砌块的夹钳形状，砂浆槽形状和竖向灰缝不采用砂浆的榫舌体系。

硅酸钙砌块　　从 1880 年取得硅酸钙砌块的专利后，人们就开始生产硅酸钙砌体。与普通砖不同，硅酸钙砌块不需烧制。然而砂子、水和石灰的混合物需在高压状态下硬化。

与黏土砖类似，根据孔隙比：上限为底面面积的 15%，将实心硅酸钙砖和成孔硅酸钙砖加以区分。两种类型的砖块高度都要小于 113mm；高度更大的称为硅酸钙砌块或空心硅酸钙砌块。对于暴露于空气中的砌体，可采用硅酸钙饰面砖和硅酸钙工程砖。在采用较薄砂浆层和 R 型硅酸钙砌块的情况下，应使用预制硅酸钙砖。R 型硅酸钙

图48：
多孔混凝土砌块

表4：
加气、轻质混凝土构件

材料：	石灰、石英砂、水泥、水和成孔膨胀剂（铝）		
骨料：			
制作：	搅拌、压模和高压下硬化		
尺寸	单位：mm（例如：240×300×23）		
砌块类型		强度等级*	密度等级*
	加气混凝土砌块	2~4	0.4~0.7
	预制加气混凝土砌块	2~4	0.4~0.7
	加气混凝土板	不承重	
	预制加气混凝土板	不承重	

注：* 普通等级

加气混凝土构件

砌块的竖缝由于采用榫舌体系，故不需使用砂浆（见"施工工艺标准"之"装修工艺标准"和"新方法"）。

此种类型的砌块于19世纪晚期开发出来。加气混凝土构件的制作：将石英砂、石灰和水泥的混合物与水搅拌，倒入模具中，并根据使用目的配置钢丝网片加强；采用铝粉做膨胀剂，通过释放氢气，增大孔隙比到材料体积的90%；未成型的材料被切割，并在高压下硬化成型。

多孔混凝土与天然矿物雪硅钙石一样，由于具有较大的孔隙比，因此具有良好的隔热和隔声性能。

当采用较薄的砂浆层时，大尺寸的加气混凝土砌块或预制加气混凝土砌块用来砌筑承重墙体。

加气混凝土板和预制加气混凝土板仅应用于不同承重体系中的非承重墙体及隔声墙体。

图49:
混凝土和轻质混凝土砌块

表5:
混凝土和轻质混凝土砖和板

材料:	矿物骨料和水硬性胶结剂		
骨料:	浮石、制作轻质混凝土的膨胀性黏土		
制作:	搅拌、压模		
尺寸	单位:mm(例如:240×300×238)和DF的倍数(如:10DF)		
砖块种类	类型	强度等级*	密度等级*
	混凝土砖		
	实心混凝土砖	12~20	1.6~2.0
	实心混凝土砖	12~20	1.6~2.0
	空心混凝土砖	2~12	0.8~1.4
	混凝土饰面砖	12~20	1.6~2.0
	混凝土饰面砌块	12~20	1.6~2.0
	轻质混凝土砖		
	轻质实心混凝土砖	2~6	0.6~2.0
	轻质实心混凝土砌块	12	1.6~2.0
	带狭槽		0.5~0.7**
	带凹槽并具有特殊隔热性能		
	预制砖		
	空心的	2~6	0.5~0.7
	轻质混凝土墙体构件	非承重	
	空心轻质混凝土墙体构件	非承重	

注: * 普通等级
** 标准规范—阐明特殊的隔热性能

作为对经典砌体结构的补充,与层高等高的构件和顶棚完善了产品表。

混凝土和轻质混凝土砌块

混凝土和轻质混凝土砌块在模具中浇筑和存储,直到达到其极限强度。两者的区别在于骨料的特性;对于轻质混凝土,只有具有多孔微观结构(主要包括天然浮石和膨胀黏土)的轻质骨料才能采用。

此处根据骨料和构件尺寸大小加以区别。实心砖的高度限制为115mm,而实心砌块的高度为175mm或238mm。两种类型的实心砌

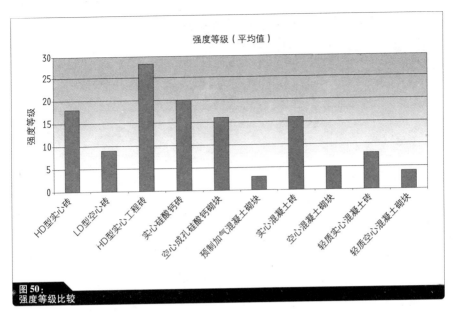

图50：
强度等级比较

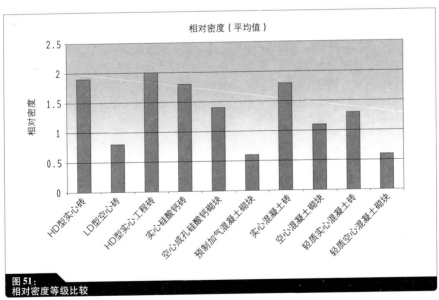

图51：
相对密度等级比较

块都不允许形成蜂窝孔，但搬运夹具口处除外。高度为238mm的空心砌块允许内部有蜂窝孔，其数量优于砌块种类（例如3K）。气候侵蚀的情况下，应采用饰面砖或饰面砌块。

表6：
砂浆类型

砂浆类型 根据 EN998-2 缩写	砂浆等级 参照 EN998-2 （CE 标记）	可用类型
普通砂浆（G）		预拌干砂浆 预拌砂浆 多室料仓砂浆 （德国建筑工地砂浆）
	M2,5	
	M5	
	M10	
	M15	
	M30	
轻质砂浆 L		预拌干砂浆 预拌砂浆 多室料仓砂浆
	M10	
	M10	
薄层砂浆 T		预拌干砂浆
	M15	

根据相同的标准对实心砖、实心砌块和空心砌块等轻质混凝土砌体单元加以区分；同时也有一些带凹槽和特殊绝缘性能的砌块和预制砖，通过尾部的 -S 或 -SW 加以识别。

制作轻质混凝土墙体结构构件和空心墙体构件以应用于非承重墙体中。

砌体砂浆类型

砂浆由胶粘剂、外加剂和添加剂构成。外加剂影响砂浆性能，诸如：防冻性能或和易性，并能很大程度上提高这些性能。添加剂通过物理过程和化学过程改变砂浆性能，但这种改变程度是有限的。添加剂包括：液化剂、缓凝剂和引气剂等。上述组分或单独供应现场搅拌使用（现场搅拌砂浆），或预先搅拌运送到现场。

重要提示：
欧洲标准没有涉及现场搅拌砂浆的内容，此处需参考德国用户手册或施工标准。

除了水外，所有其他的组分均可预先搅拌（预拌干砂浆）供应；为了节省时间，工厂预拌好的砂浆也可运送到现场使用。缓凝剂的使用考虑了砂浆必要的使用时间（预拌砂浆）。对于预拌砂浆，只有不能硬化的材料可预先拌合，而水和水泥必须在现场添加。有一种预拌砂浆可作为多室料仓砂浆供应使用。此处砂浆的组分最好现场搅拌，但却不能改变砂浆组分的配合比。

正如砌体砌块一样，砌筑砂浆可按如下详细说明：加工制作、检验、分类和性能。一体化产品标准 DIN EN 998-2 在整个欧盟均有效。

这里，将砂浆分成 3 种类型：普通砂浆（G）、轻质砂浆（L）和薄层砂浆（T）。

普通砂浆的总干密度 m 不同于轻质砂浆。普通砂浆的总干密度应不小于 $1500kg/m^3$，而轻质砂浆的总干密度小于 $1300kg/m^3$。对于标准砌块，经常采用薄层砂浆，此时砂浆层的厚度减小到 1~3mm。薄层砂浆的干密度应不小于 $1500kg/m^3$，且最大骨料粒径为 2mm。根据砂浆抗压强度的不同，将砂浆分为 M1~M30；砂浆的抗压强度值以 N/mm^2 给出。如果砂浆符合规范 DIN EN 998-2，就会采用 CE 标签加以标记。

> **重要提示：**
> 与砖块类似，在德国，德国国家标准 DIN V 20000-412（用户标准）和 DIN 18580（补充标准）均在应用；实际上规范 DIN 1053-1 也在使用。

> **重要提示：**
> 在德国，欧洲规范和 DIN 1053 关于砂浆的规定具有很大的差异性。这里，须格外注意德国用户标准和补充标准。

结 论

　　本书仅对砌体结构进行了粗浅的介绍，为读者学习砌体结构知识提供一个简单的指导。正是基于此，本书不可能包罗万象，涵盖不同国家标准的规定。不同国家标准的规定内容，应根据附录中的标准清单，分别加以论述。但本书阐述的内容对于理解砌体使用范围和施工工艺标准提供了基础，为读者进一步深入学习提供了可能。

　　应用第二章介绍的原理，我们可对多种标准结构加以研究。该原理为规划设计者阐述了一个广泛的设计方法；第三章列举的施工工艺标准更易于人们依照标准和规范实施。由相关题目，诸如混凝土结构或混凝土结构外观效果、结构性能、建筑科学高级展览会，引起的问题更易于理解。考虑建筑物将来因素进行选择时，来自制造商和经销商的信息更易因本书第四章给出的详细内容而忽略，互联网越来越成为规划设计者广泛的和重要的信息来源。

　　总之，本书为研究砌体结构的不同领域提供了基本的入门知识，并为正确掌握这些知识提供了可能。

附 录

参照标准

砌体砌块

EN 771-1（参考德国版）	砌块规范-第1部分：黏土砖
EN 771-2（参考德国版）	砌块规范-第2部分：硅酸钙砌块
EN 771-3（参考德国版）	砌块规范-第3部分：骨料混凝土砌块（密实的轻质骨料）
EN 771-4（参考德国版）	砌块规范-第4部分：蒸压加气混凝土砌块
EN 771-5（参考德国版）	砌块规范-第5部分：料石砌块
EN 771-6（参考德国版）	砌块规范-第6部分：毛石砌块

砌体砂浆

EN 998-2（参考德国版）	砌体结构砌筑用砂浆规范-第2部分：砌体砂浆

其他的建筑构件和材料

EN 845-1（参考德国版）	附加砌体构件规范-第1部分：锚件、连接构件、支撑和支架

力和荷载

EN V 1996-1-1	欧洲规范6：尺寸确定和建造砌体结构部分1-1：总则——加筋砌体和非加筋砌体标准

参考文献

Andrea Deplazes (ed.): *Constructing Architecture*, Birkhäuser Publishers, Basel 2005

Francis D.K. Ching: *Building Construction illustrated*, 3rd edition, John Wiley & Sons, 2004

Ernst Neufert, Peter Neufert: *Architects' Data*, 3rd edition, Blackwell Science, UK USA Australia 2004

Andrew Watts: *Modern Construction Roofs*, Springer, Wien New York 2006

Günter Pfeifer, Rolf Ramcke, Joachim Achtziger, Konrad Zilch: *Masonry Construction Manual*, Birkhäuser Publishers, Basel 2001

Jacques Heyman: *The Stone Skeleton: Stuctural Engineering of Masonry Architecture*, Cambridge University Press, Cambridge 1995

Theodor Hugues, Klaus Greilich, Christine Peter: *Detail Practice: Building with Large Clay Blocks and Panels*, Birkhäuser Publishers, Basel 2005

Construction Products Directive: Directive of the Council of 21 December 1988 (89/106/EEC)

Kenneth Burke: *Perspectives by Incongruity*, Indiana University Press, Bloomington 1964

Andrea Palladio: *I Quattro Libri dell' Architettura*, English translation by Robert Tavernor, MIT Press, Cambridge, Massachusetts 1997

插图致谢

P10插图：Bert Bielefeld, Nils Kummer
P34插图：Bert Bielefeld, Nils Kummer
P55插图：Gesellschaft Weltkulturgut Hansestadt Lübeck
　　　　 Willy-Brandt-Allee 19
　　　　 23554 Lübeck

图1–51:	Nils Kummer
图26:	supported by:
	Bert Bielefeld and Kalksandstein-Info GmbH (see Fig. 47)
图39, 图41	supported by:
	Deutsche Kahneisen GmbH
	Nobelstrasse 51-55
	12057 Berlin
	www.jordahl.de
图44, 图46:	supported by:
	Wienerberger Ziegelindustrie GmbH
	Oldenburger Allee 36
	30659 Hanover
	www.wienerberger.de
图47:	supported by:
	Kalksandstein-Info GmbH
	Entenfangweg 15
	30419 Hanover
	www.kalksandstein.de
图48:	supported by:
	Bundesverband Porenbetonindustrie e.V.
	Dostojewskistrasse 10
	65187 Wiesbaden
	www.bv-porenbeton.de
图49:	supported by:
	Meier Betonwerk GmbH
	Industriestrasse 3
	09236 Claussnitz/OT Diethensdorf
	www.meier-mauersteine.de

尊敬的读者：

感谢您选购我社图书！建工版图书按图书销售分类在卖场上架，共设22个一级分类及43个二级分类，根据图书销售分类选购建筑类图书会节省您的大量时间。现将建工版图书销售分类及与我社联系方式介绍给您，欢迎随时与我们联系。

★建工版图书销售分类表（见下表）。

★欢迎登陆中国建筑工业出版社网站www.cabp.com.cn，本网站为您提供建工版图书信息查询、网上留言、购书服务，并邀请您加入网上读者俱乐部。

★中国建筑工业出版社总编室
  电 话：010—58934845
  传 真：010—68321361

★中国建筑工业出版社发行部
  电 话：010—58933865
  传 真：010—68325420
  E-mail：hbw@cabp.com.cn

建工版图书销售分类表

一级分类名称（代码）	二级分类名称（代码）	一级分类名称（代码）	二级分类名称（代码）
建筑学（A）	建筑历史与理论（A10）	园林景观（G）	园林史与园林景观理论（G10）
	建筑设计（A20）		园林景观规划与设计（G20）
	建筑技术（A30）		环境艺术设计（G30）
	建筑表现·建筑制图（A40）		园林景观施工（G40）
	建筑艺术（A50）		园林植物与应用（G50）
建筑设备·建筑材料（F）	暖通空调（F10）	城乡建设·市政工程·环境工程（B）	城镇与乡（村）建设（B10）
	建筑给水排水（F20）		道路桥梁工程（B20）
	建筑电气与建筑智能化技术（F30）		市政给水排水工程（B30）
	建筑节能·建筑防火（F40）		市政供热、供燃气工程（B40）
	建筑材料（F50）		环境工程（B50）
城市规划·城市设计（P）	城市史与城市规划理论（P10）	建筑结构与岩土工程（S）	建筑结构（S10）
	城市规划与城市设计（P20）		岩土工程（S20）
室内设计·装饰装修（D）	室内设计与表现（D10）	建筑施工·设备安装技术（C）	施工技术（C10）
	家具与装饰（D20）		设备安装技术（C20）
	装修材料与施工（D30）		工程质量与安全（C30）
建筑工程经济与管理（M）	施工管理（M10）	房地产开发管理（E）	房地产开发与经营（E10）
	工程管理（M20）		物业管理（E20）
	工程监理（M30）	辞典·连续出版物（Z）	辞典（Z10）
	工程经济与造价（M40）		连续出版物（Z20）
艺术·设计（K）	艺术（K10）	旅游·其他（Q）	旅游（Q10）
	工业设计（K20）		其他（Q20）
	平面设计（K30）	土木建筑计算机应用系列（J）	
执业资格考试用书（R）		法律法规与标准规范单行本（T）	
高校教材（V）		法律法规与标准规范汇编/大全（U）	
高职高专教材（X）		培训教材（Y）	
中职中专教材（W）		电子出版物（H）	

注：建工版图书销售分类已标注于图书封底。